T0304769

Reshore Production Now

This book addresses the vital importance of reshoring US manufacturing capability to ensure economic and military security and then discusses the proven methods that the United States used to gain manufacturing supremacy in the first place. The vital takeaway is: If the job can be made sufficiently productive, the per-unit labor cost ceases to be relevant which means a business can pay high wages, realize high profits, and deliver low prices simultaneously. The contest is then not between high wages and cheap labor, but between efficiency and inefficiency and, when automation is involved, machine against machine. Readers will be able to put these principles to work very quickly to achieve tangible results.

The relatively low Federal minimum wage has meanwhile become a major issue, but inflation skyrocketed in the second quarter of 2022 when higher wages, and higher demand for goods and services, were not matched with higher productivity. The book addresses the relationship between the money supply and the velocity of money to prices, wages, and productivity.

A manufacturing resurgence in the United States will not only increase our standard of living enormously but generate taxable economic activity that will help pay down rather than increase the Federal debt. Higher productivity also delivers a greater supply of goods to accompany higher wages, and thus works against inflation. This can prevent looming recessions and disruptions.

Reshore Production Now

How to Rebuild Manufacturing and Restore High Wages, High Profits, and National Prosperity in the USA

William A. Levinson, P.E.

A PRODUCTIVITY PRESS BOOK

First published 2023
by Routledge
605 Third Avenue, New York, NY 10158

and by Routledge
4 Park Square, Milton Park, Abingdon, Oxon, OX14 4RN

Routledge is an imprint of the Taylor & Francis Group, an informa business

ISBN: 978-1-032-44540-3 (hbk)
ISBN: 978-1-032-44539-7 (pbk)
ISBN: 978-1-003-37267-7 (ebk)

DOI: 10.4324/9781003372677

Typeset in Garamond
by Deanta Global Publishing Services, Chennai, India

Contents

Preface

The 2020–2021 COVID-19 epidemic, and ensuing COVID-19 related disruptions, made clear the fragile nature of complicated international supply chains. They also underscored the enormous danger of offshoring American manufacturing capability as shown by shortages of personal protective equipment (PPE) and other critical items. The People's Republic of China (PRC) threatened openly to cut off supplies of, among other things, medications necessary to treat COVID-19 at the height of the epidemic. The Russian Federation's invasion of Ukraine in February 2022 aggravated supply chain issues even further, skyrocketing inflation compelled the Federal Reserve to raise interest rates in the second and third quarters of 2022, and the Standard & Poor 500 index is roughly, as of October 2022, 19 percent off its December 2021 high. The technical industry NASDAQ index is down almost 30 percent as of October 2022.

Gene L. Dodaro, Comptroller General of the United States and head of the US Government Accountability Office (GAO, May 5, 2022), warned, "GAO's latest report on the nation's fiscal health paints a sobering picture. Without substantive changes to revenue and spending policy, the Federal debt is poised to grow faster than the economy, a trend that is unsustainable." The GAO reference adds a figure that shows an accelerating increase in the national debt subsequent to 2020, which is essentially the opposite of what a well-run retirement account does. The idea of the latter is to put money in every year until the principal grows to the point where dividends and interest far exceed potential retirement spending. Deficit spending creates an opposite catastrophic situation in which interest on the debt will exceed any conceivable tax revenues *unless taxable economic activity expands far beyond current projections*. These developments, along with supply chain disruption due to force majeure, underscore the danger of offshoring in general and the loss of US manufacturing capability in particular.

This danger extends far beyond risks to supply chain continuity, however, because the deterioration or absence of manufacturing capability has been a *universal* leading indicator of national economic and military decline. Manufacturing is the backbone of economic and military power and a prerequisite for a middle-class living standard. Picchi (2019) reports, however, "44 percent of U.S. workers are employed in low-wage jobs that pay median annual wages of $18,000." The national minimum wage became a major political issue in 2021, but the government cannot legislate with impunity higher wages for jobs that do not create enough value to support them. Collins (2022) adds an easily foreseeable consequence of the "transition" from a manufacturing to a service economy, "But if you are among the 56% of all workers who have a high-school diploma or less, you may be struggling to make ends meet."

We have seen, in fact, that higher wages due to ordinary supply and demand issues, i.e., too many jobs chasing too few workers, result in inflation if the higher pay does not come with higher per-worker productivity. The good news is however that Henry Ford and his contemporaries proved more than 100 years ago that productivity-driven wage increases result in lower rather than higher prices, and nothing has changed since then to alter this basic principle.

The Federal debt exceeded $30 trillion in the first half of 2022 and is now greater than the United States' gross domestic product. More than a third is held by foreign interests including the PRC. Stimulus payments were necessary in 2020 to keep the badly damaged economy afloat, but further payouts of borrowed money can result only in more money chasing a limited quantity of goods and services with the obvious consequences. Higher manufacturing productivity, on the other hand, increases enormously the quantity of available goods to allow payment of higher wages, each dollar of which can buy more instead of less.

These issues—(1) the central role of manufacturing in national affluence and military power, (2) the PRC's and Russian Federation's dangerous geopolitical agendas and the PRC's record as an untrustworthy supply chain partner, (3) wages for American workers, and (4) the national deficit—therefore make reconstruction of the United States' manufacturing base imperative. The only obstacles consist of dysfunctional financial metrics and the long-discredited paradigm that cheap labor delivers low prices and high profits.

Most jobs can be made far more productive to justify higher wages and also deliver higher profits and lower prices at the same time. This is not a theory but rather a proven fact. We achieved this on a massive scale more than a century ago, we outproduced the Axis and our Allies combined during the Second World War, and nothing stops us from achieving similar results today. People such as Frederick Winslow Taylor, Frank Bunker Gilbreth, and Henry Ford proved more than 100 years ago that no amount of cheap labor, or even slave labor, can compete with high-wage efficient labor that has the support of automation.

Ford (1926, 2) wrote of this,

> we already have enough tested ideas which, put into practice, would take the world out of its sloughs and banish poverty by providing livings for all who will work. Only the old, outworn notions stand in the way of these new ideas. The world shackles itself, blinds its eyes, and then wonders why it cannot run!

This was more than a decade after he made good on these words by making the United States the wealthiest and most powerful nation on earth and creating an affluent middle class. This book's mission is to deliver the same tested ideas, which have been improved during the past 100 years by Toyota and other productivity leaders, and give readers what they need to dispel the old, outworn notions in question to achieve similar results today.

This book's first three chapters underscore the problems and also the opportunities: (1) higher wages are inflationary in the absence of higher productivity, but higher productivity enables simultaneous high wages and low prices, (2) loss or decline of manufacturing capability is a menace to our national affluence and security, and (3) the PRC's acquisition of manufacturing capability at our expense increases the associated dangers enormously. The fourth chapter shows how dysfunctional financial performance metrics and the misguided assumption that cheap labor delivers low prices and high profits created this situation in the first place. Aristotle predicted more than 2,000 years ago that automation, as then featured primarily in Greek myths and legends but also some very real inventions, would render slavery and low-wage labor obsolete. Thomas Edison projected that higher productivity would one day allow the payment of almost limitless wages, and his contemporaries such as Henry Ford proved this to be possible. The fifth chapter offers ***proven solutions*** that make labor costs irrelevant to enable the reshoring and expansion of American manufacturing capability and also encourages American consumers to demand value in exchange for their money.

This book is directed not only to manufacturing professionals but also to service industries because many of the principles are applicable to services as well. Labor leaders who want higher wages for their stakeholders will also benefit from the content, along with all citizens who seek a better understanding of the relationship between manufacturing and national prosperity. The time to regain the United States' patrimony, our inheritance from industrial leaders like Ford, Edison, Gilbreth, and Taylor, is now.

Introduction

The Preface discussed the need for the United States to regain its leadership in manufacturing to address the issues of national security, national affluence, wages, inflation, and the national debt. The next question is as to how to achieve this, and Emerson (1924, 12) answered this question more than 100 years ago in the context of the Franco-Prussian War. France had more soldiers, more manufacturing equipment, and better weapons than Prussia, but Prussia nonetheless captured Paris and overthrew the Second French Empire. Emerson explained (emphasis is mine),

> [Helmuth] von Moltke's task was however far more difficult. *He could not count on having as many men, as much money, as abundant equipment, or as much material, as his opponents.* It was evident to him that invisible theories and principles, which his self-sufficient opponents did not recognize until too late, would have to make up for meagre material resources, human lethargy, and awkward equipment.
>
> *The struggle, before it began, even in its first planning, was to be one of efficiency against inefficiency*; of efficiency, applying to the army all the twelve principles, through a new conception and shaping of military organization.

Freyberg, Conrad. 1877. Generalfeldmarschall Helmuth Graf von Moltke

Moltke therefore had to do with Prussia's Army what was later asked and required of countless factory managers as depicted by Henry Ford (Ford and Crowther, 1922, emphasis is mine): "If we only knew it, every depression is a challenge to every manufacturer to put more brains into his business—*to overcome by management what other people try to overcome by wage reduction.*" Wage reduction includes, by implication, offshoring jobs to obtain cheap foreign labor.

The issue is indeed a contest between efficiency and inefficiency, as opposed to which competitor has more workers and cheaper workers, because Frederick Winslow Taylor proved that what he called a "high-priced man" could achieve three or four times as much as a cheap one with no capital expenditures whatsoever.[1] Automation and machinery can increase the ratio to more than a thousand to one. Thomas Edison predicted roughly 100 years ago that if jobs could be made sufficiently productive, manufacturers could then pay almost unlimited wages. The hourly wage ceases to be important if it can be divided by sufficient output, and this has been achieved repeatedly since the beginning of the Industrial Revolution. Only when output does not increase commensurately with wages does inflation become a problem.

Emerson then describes how Meiji Japan adapted Moltke's principles to industrial organization to turn resource-poor Japan into a manufacturing and military powerhouse that could humiliate Imperial Russia in 1905 and terrify American industrialists with the prospect of competition from organizational systems the latter could barely imagine. The history of resource-poor Israel also has been a "struggle of efficiency against inefficiency" as its military enemies have always lost (1948, 1956, 1968, and 1973) despite overwhelming numerical superiority while its people's standard of living is unmatched in the entire region. Resource-poor modern Japan and Switzerland have similarly among the highest standards of living on earth. There is therefore no excuse for the resource-rich United States and its 330 million people to not achieve similar results, especially because it already did so in the early twentieth century.

Reshoring Is a SMART Goal

SMART stands for Specific, Measurable, Achievable, Realistic, and Timely, and reshoring is all five.

1. The specific goal is to bring back to the United States jobs that were sent offshore in a misguided quest for cheap labor and also to make jobs more productive so they can pay higher wages, reduce prices, and deliver higher profits without causing inflation. This will also generate taxable economic activity to counteract the Federal deficit.
2. Reshoring is measurable in terms of US industrial output, and general results are measurable in (for example) dollars of output or, even better, physical products per hour of labor. More value per hour of labor can be shared by all relevant interested parties or stakeholders including workers, customers, and investors. It is perhaps more to the point that a product is either "Made in the USA" or, as is the case with high-priced designer label goods made by cheap offshore labor, "Imported," and the sellers don't say from where. If the item is from a high-wage country like France or Italy that is known for good quality, the label usually says so. Price tags suitable for high-quality made-in-America products belong on high-quality made-in-America products and not on low-quality cheap products from offshore sources.
3. Some people might argue that reshoring is neither achievable nor realistic because companies cannot afford to pay high wages to American workers. History proves to the contrary that we did exactly that, and the jobs should have never been sent offshore in the first place.
4. Reshoring is timely because, while some actions might require a few years and large capital outlays, others can be performed in weeks, days, or, in the case of simple job changes to remove obvious inefficiencies, hours or minutes. Little or no money needs to be spent on the latter improvements.

We know this can be done because our country already did it and on an unprecedented and massive scale. Charles Buxton Going wrote in the preface to Arnold and Faroute's (1915) *Ford Methods and the Ford Shops* (emphasis is mine);

> Ford's success has startled the country, almost the world, financially, industrially, mechanically. It exhibits in higher degree than most persons would have thought possible the seemingly contradictory requirements of true efficiency, which are: constant increase of quality, great increase of pay to the workers, repeated reduction in cost to the consumer. And with these appears, as at once cause and effect, *an absolutely incredible enlargement of output reaching something like one hundred fold in less than ten years, and an enormous profit to the manufacturer.*

Benson (1923, 11) added, "Mr. Ford is a billionaire only because he sells automobiles and tractors for less than they were ever sold before, and pays the mechanics who make them more than mechanics were ever paid before." A billion dollars is a lot of money today, but it was inconceivably more in the money of the 1920s when the dollars were made of silver and a dime would buy a loaf of bread. The same went for the five dollars a day, and then six dollars a day, that Ford's workers received. These wages enabled one worker to maintain an entire family on a middle-class lifestyle, in contrast to many of today's two-earner families that struggle to make ends meet.

While most 100-fold and 1,000-fold productivity improvements require large capital investments in, for example, cotton harvesting machines that can do the work of more than 1000 hand laborers, a Japanese company achieved a 100-fold increase in its production of disposable gowns in response to the COVID-19 epidemic with items purchased from a 100-yen, or roughly one dollar, store (*Toyota Times Global*, 2020). This book will cite several examples of twofold to fourfold improvements for which little or no capital outlay whatsoever was necessary. In one case, a simple workplace rearrangement doubled productivity. Frederick Winslow Taylor improved the productivity of pig iron handling almost fourfold by simply having the workers pace themselves, and neither of these improvements required any expenditures whatsoever. Taylor should however have questioned the use of human labor to handle pig iron at all because conveyors were capable of doing this work in the late nineteenth century.

The benefits of reshoring should be obvious. The reconstruction of American manufacturing supremacy will not only ensure our military and economic security, it will deliver the high wages that every working American wants and ought to have without the inflation that comes from too much money chasing too few goods. These high-wage jobs will in turn address many of our society's problems including poverty in an otherwise affluent nation, and Henry Ford's industries achieved this goal long ago wherever they appeared. The resulting taxable economic activity could even reverse deficit spending to allow our country to pay down its enormous national debt, to which the need to address the COVID-19 epidemic added trillions of dollars and is still growing as of late 2022.

Content Overview

This book will begin with a discussion of the inflation that became a serious issue during the first half of 2022. This chapter was in fact added because the inflation is the worst that the United States has seen in roughly 40 years, and there are concerns it may actually lead to a recession. The book will continue with the urgent need to reshore manufacturing and go on to address the dysfunctional paradigms that led to offshoring in the first place. It will then focus on solutions that were developed more than 100 years ago to deal with competition from cheap offshore labor and were proven to work by, as but one example, the phenomenal success of the Ford Motor Company. The book will address the following topics, each with its own chapter.

1. The paradigm that a very low unemployment rate is inflationary assumes that productivity rates do not increase to support higher wages. Henry Ford proved unequivocally in the first quarter of the twentieth century that productivity enables simultaneous high wages, high profits, full employment, and lower prices. Higher productivity also offers what looks like the only way to grow taxable economic activity sufficiently to reverse the national debt spiral before it reaches the point of no return.
2. Loss or absence of manufacturing capability has always been a leading indicator of national economic and military decline. *There have been no exceptions.*
   - A sixteenth-century economist warned that Poland, which was then ostensibly one of the most powerful nations in Europe, exported commodities in exchange for manufactured goods (Zamoyski, 1987, 175). Poland was well on its way to decline even at the very moment of its military triumph at Vienna in 1683.
   - Mahan (1890) described how Spain's and Portugal's discovery of treasure in the New World ruined their manufacturing industries by allowing them to import rather than produce manufactured goods, which placed them at the mercy of their rivals England and Holland, much as our own economy is now at the mercy of the People's Republic of China. Spain's last noteworthy military achievement was at Lepanto (1571), less than a century after its treasure galleons began to bring back gold and silver from Central and South America. The same galleons proved hopelessly outclassed by the Royal Navy in 1588. Even much of the credit for Lepanto belongs to Venetian galleasses from the Venetian Arsenal, a shipyard with many characteristics of a moving assembly line.
   - A major cause of the War of Independence was Britain's policy of forcing its colonies to exchange raw materials for manufactured goods.
   - A principal cause of the American Civil War was the South's agrarian economy, which relied on exports of cotton in exchange for manufactured goods from England. Tariffs on the latter threatened to ruin the South's economy, and this played a major role in secession. Industrialization meanwhile could and should have eliminated the war's other cause by making slavery uneconomical as it had already done in most of the industrialized world. *No amount of unpaid labor or, by implication, low-wage labor, can compete with automation.* Aristotle pointed this out roughly 2,300 years ago when most machines existed solely in Greek myths, or perhaps what we might now call science fiction stories.
   - Emerson (1924) warned that the United States was exporting commodities in exchange for manufactured goods, which is the typical relationship between a colony and its colonial master. The development of motion efficiency (Frank Gilbreth), scientific management (Frederick Winslow Taylor), and mass production (Henry Ford) was meanwhile an apparent reaction to the threat of not only cheap offshore labor but also superior Japanese industrial business models.
   - American industrial output was directly responsible for the Allied victory during the Second World War, which reinforces the statement that reshoring is a SMART goal.
   - Manufacturing offers, however, a realistic path to world peace by eliminating the economic root causes of war, and also the widespread abolition of poverty through the creation of economic opportunities for everybody. Ford's industries proved the latter in the communities where they appeared.

3. The People's Republic of China[2] is a dangerous military and economic rival whose behavior toward Taiwan, Japan, Hong Kong, Australia, and the Philippines is reminiscent of the behavior of Germany, Japan, and the Soviet Union toward their neighbors during the late 1930s. While the PRC has no apparent desire to attack the United States itself, its regional ambitions could easily escalate to a threat of war and/or a threat to cripple American supply chains. It is therefore dangerous to our economic and military security to strengthen the PRC's manufacturing industries at the expense of our own or rely on the PRC for any raw materials or intermediate products including semiconductor devices and active pharmaceutical ingredients (APIs).
   - The PRC is also well known for endangering human life by knowingly selling counterfeit parts for military applications as well as adulterated and substandard pharmaceutical products, and domestic manufacturers get stuck holding the bag when injured parties file costly product liability suits.
   - Force majeure such as earthquakes, fires, droughts, and ships getting stuck in the Suez Canal can however disrupt complex international supply chains even with reputable suppliers, which is a good argument for reshoring in general.
4. Cheap labor is a dangerous and dysfunctional illusion that is unfortunately supported by cost accounting metrics that treat labor as a direct cost and even add overhead charges to the labor. Slavery along with corvée and robot (taxes in the form of labor rather than money), the ultimate forms of cheap labor, ceased to be economically viable roughly 200 years ago. The contest is therefore not one in which cheap labor enjoys an advantage over high-wage labor but instead, as Emerson (1924) put it, efficiency against inefficiency in which the high-wage labor cited by Frederick Winslow Taylor enjoys overwhelming superiority over cheap labor.
   - To put this in perspective, suppose it was possible to create artificial humans or androids that would do anything they were told and had no needs above the two lowest rungs, physiological needs and safety, of Maslow's needs hierarchy.[3] Aldous Huxley's *Brave New World* envisioned, for example, nearly mindless workers known as Deltas and Epsilons who were little more than pairs of hands and could be kept happy with cheap entertainment and a drug known as soma. Cordwainer Smith's (Colonel Paul Linebarger's) science fiction featured underpeople, or human–animal hybrids, who were created for the same purpose. The truth is that the cost of the physiological needs of, for example, 1000 of these hypothetical cotton workers—this book will discuss in detail the relative efficiencies of manual versus automated cotton harvesting—would far exceed the capital and operating costs of one cotton harvesting machine and the high wages of its operator. Draft animals like horses also have only physiological and safety needs. They require no compensation other than food, shelter, and veterinary care, but they were displaced by machines long ago for exactly the same reason. It costs a lot more to feed a thousand horses than to operate a 1,000-horsepower machine, and the latter requires sustenance only when it is working. This development has also elevated the standard of living for horses because those that are used for riding and companionship receive social interaction and even self-esteem from the human–equine relationship.
   - Low wages are usually indicators not of low prices and high profits as one might expect, but high prices and low profits instead. This is counterintuitive but low wages give enormous wastes of labor, and these can amount to 90 percent or more, a place to hide. Cheap labor has no incentive whatsoever to remove the waste because improvements will not improve the workers' situation and will lead only to layoffs. High-wage workers

and their employers know that the cost of all forms of waste comes out of their pay envelopes and profits, respectively, so they will be very diligent about the removal of all forms of waste (muda) on sight.

- Dysfunctional financial performance metrics can be extremely destructive to organizational performance because accounting methods that are mandatory for tax statements and financial reports are largely unsuitable for managerial decision-making. The total cost of ownership (TCO), or total cost of use, of a purchased item can far exceed its obvious price tag.
- Luddism is another dangerous paradigm that contends that higher productivity will result in fewer employment opportunities. Famous Luddites included the Romans and Chinese (who objected to advanced harvesting machines that would put slaves and peasants out of work), the otherwise highly capable Queen Elizabeth I and, as recently as the year 2000, the International Longshore Workers' Union (ILWU). This has never happened, and it never will because higher productivity enables lower prices, which increases the quantity demanded for the product or service in question. Henry Ford's mass production methods created hundreds of thousands of high-wage jobs in his factories along with mining and railroad transportation by making automobiles affordable to the middle class. *Luddism, in fact, destroys jobs by building waste into the price of the service or product to make it less affordable.*

5. The American worker can be (and was under Henry Ford's management system) made sufficiently productive to earn high wages side by side with high profits for the employer and low prices for the customer. Ford added that labor and management were not adversaries but should instead work together to improve productivity to make this possible. The frontline worker is often in the best position to identify opportunities of this nature.
   - Sufficiently high efficiencies make the per-unit labor cost, which is the only cost that counts in a contest between high-wage and cheap labor, negligible. There are numerous examples of 100-fold improvements and sometimes without the need for massive capital investments.
   - Labor and management must understand the all-important concept of what General Carl von Clausewitz (1976, 119) called *friction*, "the force that makes the apparently easy so difficult." This single word describes the enormous wastes and inefficiencies that are built into many if not most jobs. Its removal has resulted in improvements of up to 300 percent, and without any capital expenditures whatsoever, as proven in the nineteenth century by Frederick Winslow Taylor. *Standard work* is a valuable countermeasure to many forms of friction, and it supports job safety analysis and quality planning in the bargain.
   - Automation can, on the other hand, deliver improvements of 1000-fold or even more. One worker with a modern cotton harvesting machine can do more work than a thousand hand laborers which makes his or her wages essentially irrelevant to the cotton's price, and the same goes for countless other industries. The labor content of a Ford Model T was meanwhile about 1 percent of its total price, and even the labor content of modern vehicles comes to only about 10 percent of the price.
   - The American consumer must be educated to demand value for his or her money, and celebrity endorsements and fancy designer labels on cheaply made and often poor-quality goods are not value. There should be zero tolerance for made-in-America price tags on shoddy goods labeled only as "Imported" because the seller is ashamed to admit that they were made by cheap offshore labor.[4]

Notes

1. The workforce was at the time predominantly male, and we would say "high-priced people" or "high-wage labor" today.
2. This book will use "People's Republic of China" or "PRC" as opposed to "China" or "the Chinese" to avoid confusion with Taiwan, Hong Kong, and ordinary people under PRC rule.
3. Maslow's needs are, from lowest to highest, (1) physiological needs, (2) safety, (3) social interaction, (4) self-esteem, and (5) self-actualization.
4. This does not mean the sellers misrepresent the products of cheap offshore labor as being made in the United States but rather that the prices charged are appropriate for high-quality American-made products as opposed to the shoddy goods in question.

Chapter 1

Wages, Productivity, and Inflation

There is a widespread and dysfunctional belief that a low unemployment rate is inflationary because the supply and demand model drives higher wages and therefore higher prices. The past few years proved this true to some degree when retail stores began to advertise entry-level jobs for $13 and then $15 an hour, and other companies advertised entry-level jobs for up to $20 an hour. None of these jobs required a college degree or skill in a trade such as welding, plumbing, and carpentry; these trades command extremely high wages, and rightly so. This chapter will show however that *industrial productivity not only counteracts inflation but also generates taxable economic activity that will help pay off the ever-expanding Federal debt.*

Inflation is the easily foreseeable consequence of higher wages that are not accompanied by higher productivity. Henry Ford (*Ford Ideals*, 1922, 76)[1] wrote however of the purported desirability of substantial unemployment,

> There are those who claim that a certain proportion of unemployed men is desirable from the industrial standpoint. A crowd of men clamoring around the factory gates for jobs helps keep the men inside steady and helps keep wages down, they say.
>
> This is a detestable philosophy. It is cold speculation in flesh and blood and anxiety and hunger. We don't want any condition that is dependent on unemployment for steadiness.

If high wages result from a shortage of workers rather than an increase in productivity, then inflation will indeed be an obvious consequence. Another way to say this is that a certain amount of unemployment is a necessary evil *in the absence of higher productivity*, which makes those who view low unemployment as problematic partly right. This attitude prevails to this very day because "Yet that increase [in the unemployment rate from 3.5% to 3.7%] was also an encouraging sign: It reflected a long-awaited rise in the number of Americans who are looking for work" (Rugaber, September 2, 2022). Lash et al. (2022) add that the stock market plunged on October 7, 2022, because "a solid jobs report for September increased the likelihood the Federal Reserve will barrel ahead with an interest rate hiking campaign many investors fear will push the U.S. economy

DOI: 10.4324/9781003372677-1

into a recession." This reference quoted Joseph LaVorgna of SMBC Nikko Securities to the effect that once the unemployment rate does start to rise, it may do so far more rapidly than desired. The purportedly desirable higher unemployment rate reduces the ability of people to buy things, which may well force prices down but will also generate more unemployment. It is easy to see why investors regard these interest rate hikes as dangerous.

Rugaber (September 20, 2022) adds of a subsequent increase in the interest rate, "The Fed intends those higher borrowing costs to slow growth by cooling off a still-robust job market to cap wage growth and other inflation pressures" but also warns that this could lead to a recession instead of the desired "soft landing."

The adage that, if the only tool you have is a hammer, everything starts to look like a nail therefore seems applicable to the Federal Funds Rate as well. This underscores the limits of the Fed's ability to manage inflation. If higher prices are due to genuine shortages that are aggravated by the kind of supply chain failures that result from offshoring of key raw materials, intermediates, and finished goods, higher interest rates will do little to counteract the basic economic laws of supply and demand. If some people react to the prices by doing without, then the sellers will have to lay off employees, and the result will be an economic contraction. These conditions are however also an enormous opportunity for sellers who can increase their output to meet the unfilled demand at reasonable prices.

Stimulus payments with borrowed money might have been a necessary response to the COVID-19 epidemic in 2020, when business closures threatened an economic catastrophe, but their long-term effect also was easily foreseeable. Ford (*Ford Ideals*, 1922, 177) added,

> The demand of the disorderly element is practically that everybody be requested to raise fewer potatoes, and yet that everybody be given more potatoes ... If everybody does less work and everybody gets more of the product of work, how long can it last?

The reference to potatoes is very instructive because they represent genuine value rather than money, which is nothing more than a medium of exchange.

Money Is Not Value or Utility

The first step is to recognize that money is not value or utility. Economists sometimes use the word "util" as a hypothetical measurement of utility or what customers can actually do with a product or service. Money does not have utility; it is only a medium of exchange for utility. We can't eat money, but we can exchange it for potatoes, which we can eat. The potato farmer accepts money because he or she can, in turn, exchange it for something else like a potato digger.

There was in fact a time in human history when money did not even exist, and people bartered utils for utils. Even during the Great Depression, when money was simply not available to many people, farmers would (for example) barter food in exchange of veterinary care for their livestock. It is quite likely that inflation as we know it was unheard of in barter economies, although rates of exchange for individual goods and services probably did fluctuate with supply and demand.

Barter requires, however, that the parties to a transaction come up with an agreement for every transaction, such as potatoes for eggs, clothing for horses, and so on. Money facilitated transactions by allowing sellers to set prices without reference to any underlying commodity. Some societies used valuable commodities, such as bricks of tea in China and tobacco in colonial Virginia, Maryland, and North Carolina, as money. Gold, silver, and copper became popular coinage because their relative scarcity limited the money supply, but this did not eliminate all the underlying issues. The debasement of money, whether by incorporation of base metals long ago or inflation today, is the most dangerous.

Money Debasement, Then and Now

Mannix (1958, 3–4) reports that Imperial Rome imposed tariffs to protect Roman workers from cheap foreign labor, but the tariffs, however, made it impossible for foreign nations to sell their products to Rome.

> the [Roman] government was finally forced to subsidize the Roman working class to make up the difference between their "real wages" (the actual value of what they were producing) and the wages required to keep up their relatively high standard of living.[2]

The problem was and is, of course, that higher wages cannot purchase more utils if nobody is producing them. Protectionist tariffs are meanwhile not the right way to keep out products of cheap foreign labor; we must increase the productivity of our own workers instead.

This idea persists almost 2,000 years later with proposals such as a "universal basic income" among American politicians and the "Basic Living Stipend" in David Weber's Honor Harrington science fiction series. Harrington is essentially C.S. Forester's Horatio Hornblower with a spaceship instead of a frigate, and the Basic Living Stipend is why Harrington's enemies in the People's Republic of Haven must wage war on their neighbors, much as the Romans did long ago, to get enough money to keep the Dolists (people on the dole) happy. Weber's Dolists seem to correspond very closely, in fact, to the Romans who decided to stop working once they got a guaranteed income from their government. Rudyard Kipling (1919) wrote of this mentality in "The Gods of the Copybook Headings,"

> In the Carboniferous Epoch we were promised abundance for all,
> By robbing selected Peter to pay for collective Paul;
> But, though we had plenty of money, there was nothing our money could buy,
> And the Gods of the Copybook Headings said: "If you don't work you die."

This underscores a key takeaway from this chapter; *we cannot pay somebody more than what he or she creates without causing inflation*. Mannix goes on to describe the economic havoc that resulted from these practices. Roman Emperors had to pay meanwhile for gladiator fights and also human–animal fights that came close to driving many African animals to near extinction. The less responsible ones such as Caligula and Nero spent enormous sums on themselves as well, and Rome simply did not produce enough value (or utils) to support these expenditures. Desjardins (2016) describes how the Emperors debased their money by reducing the silver content: "By the time of Marcus Aurelius, the denarius was only about 75% silver … By the time of Gallienus, the coins had barely 5% silver. Each coin was a bronze core with a thin coating of silver."

The latter is in fact reminiscent of the "clad" quarters and half dollars with which the United States replaced its silver coins in the 1960s, although the United States no longer pretended at that point that its money had any intrinsic value. Romans, however, had to rely on the proposition that the denarius had intrinsic value so the consequences of this debasement, as depicted by Desjardins, were that "Hyperinflation, soaring taxes, and worthless money created a trifecta that dissolved much of Rome's trade." The subsequent collapse of Rome was due primarily to economic dysfunction rather than any barbarian military superiority. Carl Icahn, chair of Icahn Enterprises, compared the situation in Rome, which experienced 15,000 percent inflation between the years 200 and 300, to that in the United States in the second half of 2022 (Daniel, 2022).

Many countries recognized, however, the kind of economic havoc that could come from debasement of money, and England defined as treason the act of clipping coins (shaving silver away from the edges for resale) along with counterfeiting. There is in fact a relationship between treason and debasement, noting that counterfeiting has been used as a weapon during wartime to destroy confidence in the enemy's money and thus undermine the enemy's economy. Germany, for example, counterfeited British money during the Second World War, the Union circulated fake Confederate money during the American Civil War, and the British circulated phony American money during the War of Independence (Bickford, 2012).

Even Precious Metals Often Lack Genuine Utility

Suppose that Rome had, instead of debasing its currency, discovered gold and silver in the New World, or Mali in Africa, and issued more denarii with the proper silver content. Inflation would have still resulted because more money would have been chasing the same number of utils or units of actual value and utility. This is not mere speculation because when Mansa Musa, the King of Mali, visited Egypt, he spent and even gave away so much gold that the value of gold plummeted (*National Geographic*, no date given). This book will show later how Spain and Portugal mistook the gold and silver they discovered in the New World for "utils," traded the gold and silver to their rivals England and Holland for real "utils," and ruined their own manufacturing capability as a result.

While gold, silver, and copper have numerous industrial uses, as do lithium (for batteries) and rare earths, their real value consists solely of what one can do with them. It is to be remembered that King Midas starved to death because everything he touched, including his food, turned into gold. This is why I am not particularly receptive to radio ads that advertise gold as an investment. While gold is admittedly a hedge against inflation, it produces no value or utility unlike a stock backed up by an actual factory, container ship fleet, chemical plant, or similar producing asset. It should be regarded instead as a commodity whose actual value depends on its industrial demand. Cryptocurrency may be even worse because, unlike gold, it has no intrinsic value whatsoever.

Weimar Wastepaper and Cryptocurrency

The Weimar Deutschmark was, along with the Zimbabwe dollar, the epitome of hyperinflation because pieces of paper with 10-figure or even 13-figure numbers on them have even less intrinsic value than coins regardless of debasement of the latter. The modern penny, which is primarily zinc, is at least worth its weight in zinc. Germans did, however, use Weimar Deutschmarks for wallpaper, kites, tinder for lighting stoves (*Rare Historical Photos*, no date given), and even allegedly as toilet paper. The only value of cryptocurrency is meanwhile its utility for anonymous transactions, and it is otherwise purely speculative like seventeenth-century Dutch tulip bulbs and the dot-com stocks of 2000–2001. Royal (2022) reports that Bitcoin is off 72 percent from its 2021 high as of September 2022, and Ethereum is down 73 percent. The performance of blockchain computations for the sake of performing blockchain calculations, and this is how cryptocurrency is "mined," has negative utility because it uses enormous quantities of electricity along with the capital costs of the computers involved but produces nothing of actual value.

Money Supply and Velocity, and the Equation of Exchange

Labor is, like anything else of value, subject to supply and demand models. A shortage of labor will require employers to increase wages which will, in the absence of commensurate productivity, be inflationary even without currency debasement. The *velocity of money* is the number of times inside a given time period that a unit of currency changes hands. This can result in higher wages, and wages are essentially the price of labor, regardless of the money supply. Wen and Arias (2014) offer the *quantity theory of money* which says (Equation 1.1):

$$M \times V = P \times Q \tag{1.1}$$

where
- M = money supply
- V = velocity of money
- P = price
- Q = quantity of goods and services or "utils"

This is also known as the *equation of exchange* (Mishkin, 1986, 401) and is sometimes known as the Fisher equation after Irving Fisher, the author of *The Purchasing Power of Money.* Fisher (1922, 20) explains,

> If the number of dollars in a country is 5,000,000, and their velocity of circulation is twenty times per year, then the total amount of money changing hands (for goods) per year is 5,000,000 times twenty, or $100,000,000. This is the money side of the equation of exchange. Since the money side of the equation is $100,000,000, the goods side must be the same. For if $100,000,000 has been spent for goods in the course of the year, then $100,000,000 worth of goods must have been sold in that year.

We can express this as a summation (Equation 1.2), where p_i is the price of the ith service or product, and q_i is the quantity exchanged. Note for future reference that labor is itself a service with an associated price.

$$M \times V = \sum p_i q_i \tag{1.2}$$

Fisher (1922, 20) uses as an example 200 million loaves of bread @0.10, and a dime would indeed buy a large loaf of bread in 1922, to account for $20 million of the $100 million on the left side of the equation. Add 10 million tons of coal @$5.00 for $50 million and 30 million yards of cloth @$1.00 for $30 million to balance the equation. Fisher (1922, 23) even uses a balance to illustrate this as shown in Figure 1.1. The money supply is a weight on the left side of the fulcrum, and the velocity of money is its position; more velocity gives the money supply more leverage. The quantities of bread, coal, and cloth are the weights on the right side, and their positions are their prices.

It is accordingly clear that an increase in the velocity of money in the presence of a constant money supply, or an increase in the money supply (such as debasement) in the presence of a constant velocity, will result in higher prices *if the quantity of goods and services remains constant.* If we return to Spain's and Portugal's discovery of "money" in the New World, the effect was to increase

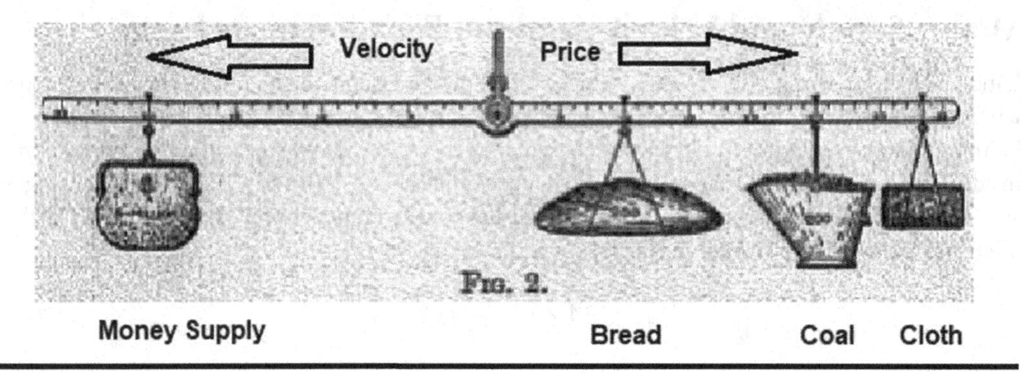

Figure 1.1 Equation of Exchange as a Balance

their money supplies enormously without any corresponding increase in production, so the consequences should have been easily foreseeable. Fisher (1922, 21) explains quite explicitly,

> Suppose, for instance, that the quantity of money were doubled, while its velocity of circulation *and the quantities of goods exchanged remained the same* [emphasis is mine]. Then it would be quite impossible for prices to remain unchanged. The money side would now be $10,000,000 × 20 times a year or $200,000,000; whereas, if prices should not change, the goods would remain $100,000,000, and the equation would be violated. Since exchanges, individually and collectively, always involve an equivalent quid pro quo, the two sides must be equal. Not only must purchases and sales be equal in amount—since every article bought by one person is necessarily sold by another—but the total value of goods sold must equal the total amount of money exchanged. Therefore, under the given conditions, prices must change in such a way as to raise the goods side from $100,000,000 to $200,000,000. This doubling may be accomplished by an even or uneven rise in prices, but some sort of a rise of prices there must be.

We have seen so far that an increase in the product of the money supply and its velocity will result in higher prices, and therefore inflation, *in the absence of commensurate increases in productivity.* This chapter will show, however, that higher productivity changes everything.

Deficit Spending and Inflation

The stock market experienced a major decline in the first three quarters of 2022 when the Federal Reserve raised interest rates to counteract inflation and another decline in September. Mishkin (1986, 618–619) warns however that deficit spending promotes inflationary monetary policy regardless of what the Federal Reserve does and adds that anti-inflationary action by the Fed loses credibility in the absence of restraint by the government. He (620) describes how both Ronald Reagan and Margaret Thatcher engaged in deficit spending while they attempted to pursue anti-inflationary policies, and the result was an increase in unemployment in the United States and the United Kingdom.

The same phenomenon occurred in 2020 and 2021, when the Trump and Biden Administrations issued stimulus payments (backed with borrowed money) to avert an economic collapse due to the COVID-19 epidemic. Generous, and perhaps overly generous, unemployment compensation payments caused workers to stay out of the workforce even when it was safe to go back, with the easily foreseeable result that employers had to pay higher wages for the same amount of productivity. Inflation is the obvious consequence of this, we started to get it in 2021, and we have a bear stock market to date (October 2022). The next potential consequence might be an actual recession, and this is why investors are getting out of the stock market.

The Federal Reserve's only weapon against inflation consists of its ability to regulate the money supply via the discount rate (which it has raised several times in 2022) and by selling or buying debt, but industrial productivity can also suppress inflation regardless of the money supply and/or velocity. Industrial productivity can in fact increase the velocity of money enormously, which increases taxable economic activity to counteract the deficit.

Productivity Counteracts Inflation and Pays Down the Deficit

If we note that labor is in fact a service, then the price of labor per util produced will increase side by side with the price of utils if the quantity of the latter is held constant. Suppose we rewrite the quantity of money theory as shown in Equation 1.3, which will disregard for convenience other transactions that involve profits, materials, energy, and capital costs related to production. These considerations will be addressed later, but they do not change the underlying conclusions. We will assume that labor is the only cost. Revenue is the price of utils times the quantity produced and sold, and wages are the product of the cost of labor (e.g., in dollars per worker) and the number of workers.

$$M \times V = \text{Revenue} + \text{Wages} \quad (1.3)$$

There are four vital takeaways from Equation 1.3.

(1) *Useful goods and services, or utils, are the only items of genuine value in the equation.* Wages are of value only if they produce utils; wages that are wasted on non-value-adding activities are of no value to customers, employers, or even the workers themselves in the long run because wages cannot buy utils that are not produced. *This makes wasted labor inflationary*, and the same goes for all other forms of waste.

(2) As revenue must balance wages, then any increase in wages will by necessity increase prices if the number of utils produced remains constant. Another way of saying this is that *we cannot get something (higher wages) for nothing (no increase in productivity)* or, as Henry Ford put it, "The demand of the disorderly element is practically that everybody be requested to raise fewer potatoes, and yet that everybody be given more potatoes."

(3) *If $M \times V$ is constant but util production increases substantially, then per-worker wages can increase while per-util prices decrease.* This is exactly how Ford increased wages, profits, and automotive sector employment while he lowered the price of the Model T. This takeaway also eliminates the arguments of the Luddites, whom this book will discuss in further detail later, that higher productivity reduces employment.
   - Fisher (1922, 22) adds explicitly, "a doubling in the quantities of goods exchanged will not double, but halve, the height of the price level, provided the quantity of money and its velocity of circulation remain the same." This assumes however that none of the benefits of higher productivity are shared with the workers in the form of higher wages and the employer as higher profits. Even under these conditions, however, price reductions significantly less than 50 percent, but significantly more than zero, will result in a higher quantity demanded to keep the workers employed with the higher wages in question.

(4) *Not only can higher productivity negate any inflationary effects from an increase in the velocity of money, this increase is a taxable economic activity that counteracts the deficit.* We will see that MV can easily rise side by side with declining prices due to the greater quantity of utils, and MV often represents taxable transactions.

Revenue Must Also Balance Profits and Costs of Production

The next step is to expand Fisher's balance illustration to show that *there is a second balance between utils on one side and labor, materials, and profits on the other.* We can add cost of capital, energy, and fixed operating expenses to the right side of this balance, and these costs deserve attention, but we will focus for simplicity only on labor, materials, and profits.

The first balance (Figure 1.2) shows that MV, the money supply times velocity, must equal the weighted (by cost) sum of utils produced along with the labor cost, material cost, and profit per util. Suppose for example that:

- The money supply M is 200 in arbitrary units, and the velocity is 12. The weight of MV on the left side of the balance is therefore 2,400.
- Production is 200 utils for which:
  - The profit is 1 per util, for 200.
  - The material cost is 2 per util, for 400.
  - The sale price is 6 per util, for 1,200.
  - Fifty workers are paid 12 each, for 600.
- The total on the right side of the balance also is 2,400 because the money that is circulated must equal the profit, material cost, sales revenue, and labor cost.

We can add profits and material costs to Equation 1.3 to get Equation 1.4, noting that the sales revenue equals the quantity of utils times their sales price. If revenue, profits, labor costs, and material costs increase due to higher productivity rather than higher prices, then MV can also increase without inflation—and MV is taxable economic activity.

$$M \times V = \text{Profits} + \text{Material Costs} + \text{Labor Costs} + \text{Sales Revenue} \quad (1.4)$$

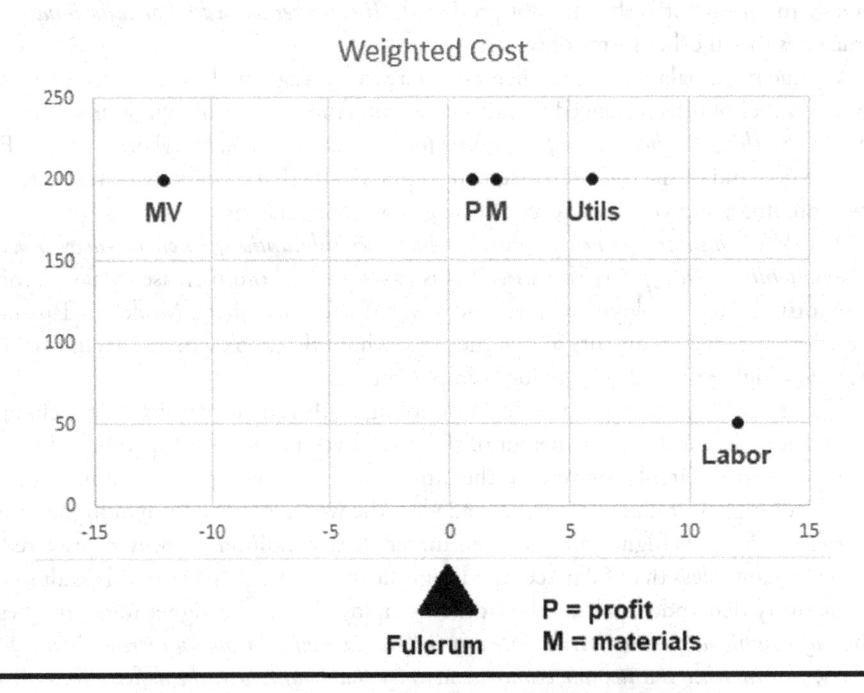

Figure 1.2 M × V Must Balance Revenue, Materials, Labor, and Profits

The second balance (Figure 1.3) shows that the revenue from the utils must balance the labor cost, material cost, and profit per util. In this case, the revenue of 1,200 monetary units must cover the profit of 200@1, material cost of 200@2, and labor of 50@12.

We can express this balance as Equation 1.5, which shows why many decision-makers think they can reduce prices and/or increase profits by offshoring jobs for cheap labor.

$$\text{Sales Revenue} = \text{Profits} + \text{Material Costs} + \text{Labor Costs} \tag{1.5}$$

The short-sighted approach, or what Emerson (1924, numerous pages) calls "near common sense," is to try to reduce the labor cost by cutting wages or offshoring jobs. The intelligent approach, which Emerson calls "supernal common sense" and was put into practice by Ford, Taylor, Emerson, and their contemporaries, is to increase the number of utils produced by each worker. Figure 1.4 shows what can happen if per-worker productivity doubles, but without any change in material cost. Note that we still have 50 workers as shown by the ordinate, but each is now twice as productive and receives higher wages as shown by the position on the axis. The price-weighted quantity of utils is the sales revenue, which must balance all costs related to production, but the price reflected by the position on the axis is now lower.

- Production is 400 utils for which:
  - The revenue is 400@5 = 2,000; note that the price has been lowered from 6 to 5 to help increase the quantity demanded and thus keep all 50 workers employed at the higher output.
  - The profit is 400@0.75 = 300. Even though the producer gets 25 percent less per util, it makes twice as many so its profit is 50 percent greater than it received for 200@1.

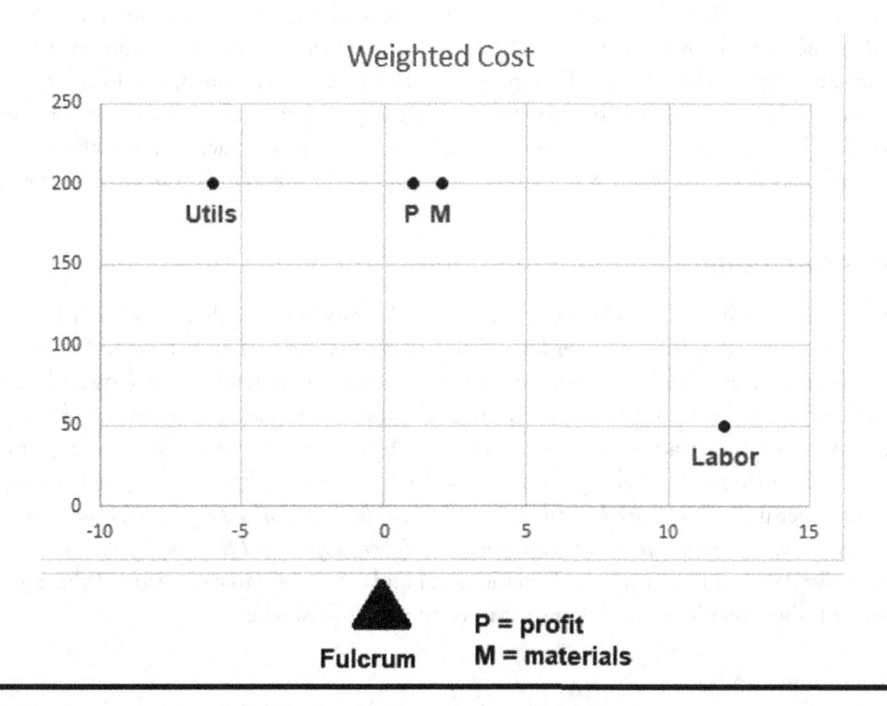

Figure 1.3 Revenue Must Balance Production Costs and Profit

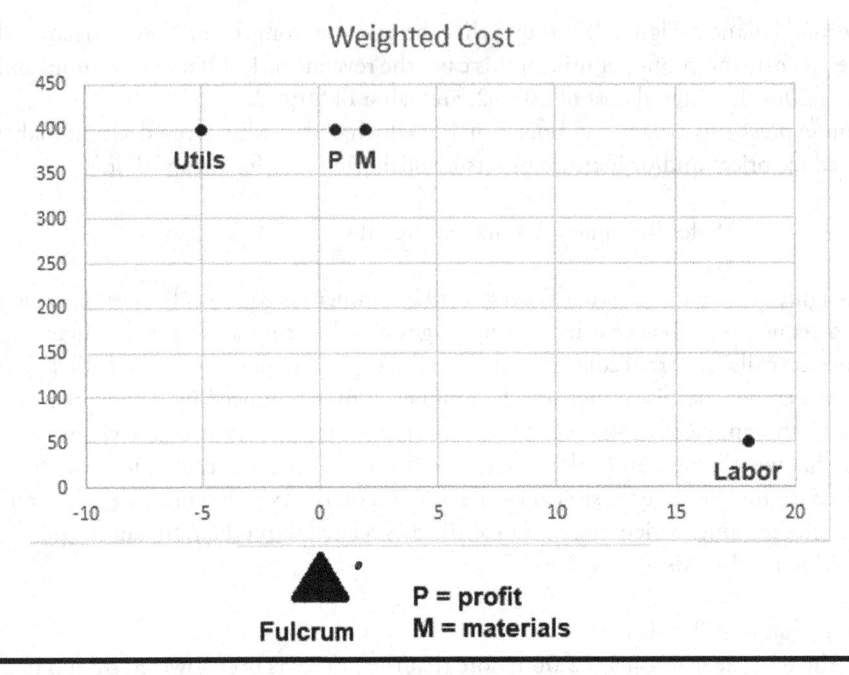

Figure 1.4 Effect of a Productivity Increase on Wages, Prices, and Profits

- The material cost is 400@2 = 800 as the material still costs 2 per util, but we will see later that attention to wastes of stock and consumables can reduce this cost as well.
- Fifty workers are paid 18 each for 900; their wages are now 50 percent higher.

■ MV is now 4,000 rather than the previous 2,400 which, in the absence of any change in the money supply, represents a 66.7 percent increase in the velocity of money. One would normally expect this to be inflationary, but the price of the output is lower rather than higher per unit. Higher productivity can therefore negate inflationary pressures entirely. This higher velocity of money also represents taxable economic activity, which suggests that *the only way to get out of our $30 trillion (and climbing) Federal debt is to produce our way out.*

Waste Is Inflationary

The previous sections have shown how wasted labor adds costs but no utility, which is inflationary by definition. The same goes for any other form of waste because its cost must be reflected in the price of the goods or services. Henry Ford paid attention to the cost of material and energy and stated that there are in fact only three kinds of waste: waste of time, waste of material, and waste of energy. Any reduction in any of these wastes reduces the height of the corresponding vertical bar (e.g., if we have separate ones for materials and energy) which enables simultaneous lower prices, higher wages, and higher profits per util. *This book's key deliverable rests on the very simple proposition that eradication of waste from any supply chain can deliver higher wages, higher profits, and full employment with no inflation whatsoever.* We can in fact rewrite Equations 1.4 and 1.5 as 1.6 (an expansion of the equation of exchange or Fisher equation) and 1.7, respectively, to reflect this, where

- R = revenue from sales
- P = profit

- M = cost of materials and energy
- C = cost of capital such as plant and equipment
- L = cost of labor
- Waste includes not only the Toyota production system's Seven Wastes but anything else that does not add value. The following wastes incorporate those depicted by Ford (time, material, and energy), while inventory is among Toyota's Seven Wastes.
  - Waste of materials, which the ISO 14001:2015 standard for environmental management systems can address if users apply it to all material wastes rather than just environmental aspects.
  - Waste of energy, the focus of the ISO 50001:2018 standard for energy management systems.
  - Waste of labor, and this book will show that this is still enormous in many occupations; this corresponds to waste of the time of people.
  - Waste associated with cycle time and inventory, which is proportional to cycle time according to Little's Law. This corresponds to waste of the time of things.
  - Transportation and handling, such as the cost of bringing goods from the People's Republic of China (PRC) to the United States. This wastes energy and lead time, which again results in inventory.

$$M \times V = R + P + M + C + L + \text{Waste} \tag{1.6}$$

$$R = P + M + C + L + \text{Waste} \tag{1.7}$$

Equation 1.6 reinforces the conclusion that *waste is inflationary* because, if the price of actual value or utility remains constant, the money supply and/or velocity must increase to cover the waste. Equation 1.7 reinforces this conclusion even further because sales revenue must cover the cost of the waste, and this means higher (inflationary) prices.

We Must Produce Our Way Out of the National Debt

The Preface included the Government Accountability Office's (GAO's) projection of spiraling Federal debt, and similar projections are just as disturbing as those of the Ghost of Christmas Future in Charles Dickens's *A Christmas Carol.* Figure 1.5 depicts a model with the following assumptions:

- The gross domestic product (GDP) was $23 trillion in 2021, and the GDP growth rate was 5.7 percent. The model assumes the growth rate will remain at 5.7 percent although it was less in 2022. The government collected $4 trillion in taxes in 2021, so the tax rate is 17.6 percent of GDP. The model assumes that this rate will remain constant in the future.
- The Federal debt was $28.4 trillion in 2021 and is closer to $31 trillion as of September 2022.
- Interest on the Federal debt is 3.0 percent, which was in fact close to the rate for one-year Treasury bills in mid-2022; it is higher as of September. The Government Accountability Office reference projects, however, that the interest rate will be closer to 4.6 percent which is consistent with the current trend.
- The Federal budget (other than interest on the debt) was $6 trillion in 2021.

The deficit was accordingly \$4.0 trillion in revenue less \$6.0 trillion for government programs and also 3 percent of \$28 trillion, or \$0.85 trillion, for \$2.85 trillion in deficit. This increases the debt to \$31.2 trillion for 2022. Figure 1.5 projects runaway debt, which is an obvious consequence of spending more money than the government brings in from taxes. The deficit adds to the debt, and the debt adds to the annual interest payments necessary to carry it.

If we use the 4.6 percent interest rate as predicted by the GAO, Figure 1.6 shows a national debt of roughly \$150 trillion in 2040 which is comparable to what the GAO got.

The only way to even begin to get out of this kind of debt is to debase our money to pay it off with inflated dollars, which will harm savers and investors while it drives interest rates even higher. Figure 1.7 summarizes this projection quite accurately.

The question is therefore what we need to do about this before we reach the point of no return. As stated by Dickens (1843),

> "Before I draw nearer to that stone to which you point," said Scrooge, "answer me one question. Are these the shadows of the things that Will be, or are they shadows of the things that May be only?"
>
> Still the Ghost pointed downward to the grave by which it stood.
>
> "Men's courses will foreshadow certain ends, to which, if persevered in, they must lead," said Scrooge. "But if the courses be departed from, the ends will change. Say it is thus with what you show me!"

The answer is yes, we can change this course away from economic disaster to national prosperity. Suppose we retain the 4.6 percent interest rate projection, but assume a 10 percent GDP growth rate. Figure 1.8 shows that the deficit becomes a surplus in the mid-2030s, and the national

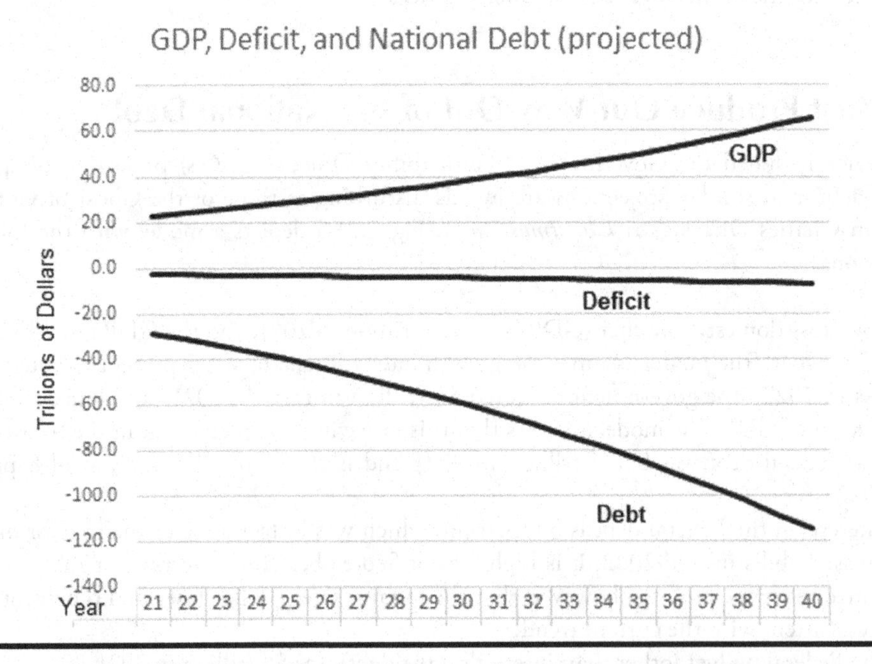

Figure 1.5 GDP, Deficit, and National Debt, Assuming 3 Percent Interest Rate

debt begins to shrink rather than expand. The national debt will in fact cease to exist circa 2045 under these conditions.

We cannot have a 10 percent annual GDP growth rate by merely wishing for it, but American industrialists such as Frederick Winslow Taylor, Frank Gilbreth, and Henry Ford proved more than a hundred years ago that the productivity of many if not most jobs can be increased by

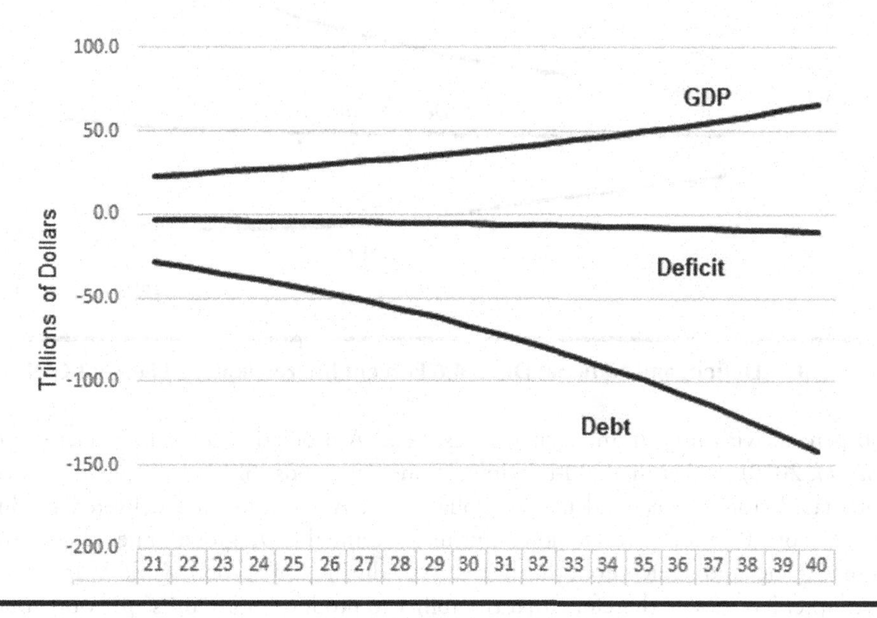

Figure 1.6 GDP, Deficit, and National Debt, 4.6 Percent Interest Rate

Figure 1.7 Ghost of Christmas Yet to Come (Arthur Rackham, 1915)

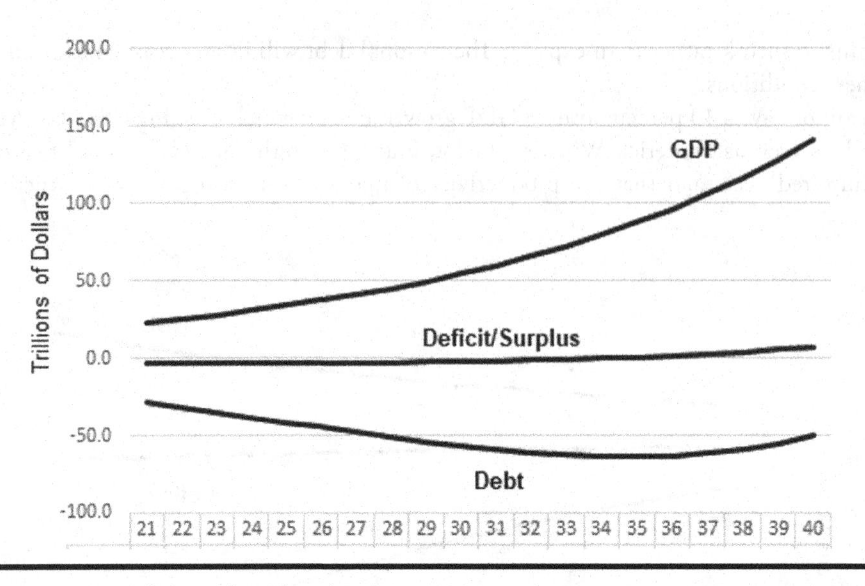

Figure 1.8 GDP, Deficit, and National Debt, 4.6 Percent Interest Rate, 10 Percent GDP Growth

100–300 percent with little if any capital investment. A protective gown manufacturer (*Toyota Times Global*, 2020) proved that 100-fold improvements are possible, at least in a few cases, with items purchased from the equivalent of a dollar store. Automation has delivered productivity improvements of 1,000-fold or even more. This makes annual GDP growth rates of ten percent or even more realistic, and we achieved this by necessity during the Second World War.

This chapter has shown, therefore, that the ability to produce more "utils" per worker not only negates both inflation and unemployment; it creates more taxable economic activity. This underscores further the desirability of reshoring our manufacturing capability. The next chapter will address the enormous dangers associated with the loss of manufacturing jobs.

Notes

1. The citation of *Ford Ideals* distinguishes this reference from Ford and Crowther's *My Life and Work*, also from 1922.
2. Mannix's *The Way of the Gladiator* was apparently the inspiration for the movie starring Russell Crowe, and it includes some humorous anecdotes as well as violence. A famous venator (gladiator who fought wild animals) named Carpophorus trained his dog to wag its tail, or whine piteously, in response to hand signals. He would then signal the dog to wag its tail when he mentioned the chariot team favored by the person to whom he was talking, and to whine when he named a rival team. The word *faction* once referred to chariot teams, so Carpophorus's dog probably gained him a lot of friends.

Chapter 2

Loss of Manufacturing Equals National Decline

The first chapter related the money supply and velocity (MV) via Fisher's equation, or the equation of exchange, to prices and wages to show that higher wages are inflationary in the absence of commensurately higher productivity. It also showed however that higher productivity can increase MV, and therefore taxable economic activity, without causing inflation, and the same goes for the removal of all forms of waste from a supply chain. This chapter will address the enormous drawbacks of offshoring and will show why it should have never happened in the first place. It will also show that, while manufacturing is the backbone of military power, it also offers an end to the root cause of war by making cooperation far more profitable than conflict.

The absence or decline of manufacturing capability, i.e., the ability to add value to raw materials and commodities, is a *universal indicator of national decline*. There have been no exceptions ever since manufacturing began to displace agriculture as the chief source of wealth. *The primary characteristic of this dangerous situation is the exchange of raw materials and commodities for manufactured goods or, even worse, the exchange of borrowed money for manufactured goods*. This is the usual trade relationship between a colony and its colonial master. Emerson (1924, 94) cited very clearly the dangers associated with the de facto export of the soil of the United States for manufactured goods.

> we childishly squander our national resources in exchange for perishable luxuries supplied us by older and wiser men, corporations, and nations, who, not having gifts and prodigal equipment, still use their brains and hands—men who trade us sunshine, water, and air for our mined wealth, for our soil's fertility.
>
> At the present market price of nitrogen, phosphorus, and potash, every pound of cotton that leaves our shores carries with it about $0.03 of soil value, every bushel of corn or wheat carries away about $0.20 of soil fertility. The nominal profit, about $0.03 a pound on cotton, about $0.20 a bushel on grain, is no greater than the market price of what is taken from soil value, and our agriculturist is devoting his great activity, his strenuous life of long hours, to the spending of his capital. The net income is nil.

DOI: 10.4324/9781003372677-2

Another way to put this is that most agricultural products are commodities, and there is a limit as to how much value one can add to a commodity. This book will show, however, that new approaches to agriculture that include fully enclosed vertical farms are likely to change this. They require less resources than traditional farms and can also reduce many of the costs of transportation of the product.

Taylor (1911) raised similarly the issue of what we now call sustainability, which recognizes the fact that natural resources are not inexhaustible.

> President [Theodore] Roosevelt in his address to the Governors at the White House, prophetically remarked that "The conservation of our national resources is only preliminary to the larger question of national efficiency."
>
> The whole country at once recognized the importance of conserving our material resources and a large movement has been started which will be effective in accomplishing this object. As yet, however, we have but vaguely appreciated the importance of "the larger question of increasing our national efficiency."
>
> We can see our forests vanishing, our water-powers going to waste, our soil being carried by floods into the sea; and the end of our coal and our iron is in sight. But our larger wastes of human effort, which go on every day through such of our acts as are blundering, ill-directed, or inefficient, and which Mr. Roosevelt refers to as a, lack of "national efficiency," are less visible, less tangible, and are but vaguely appreciated.

This suggests that our current practice of exporting raw materials, recyclables, and, even worse, borrowed money in exchange for finished goods is extremely dangerous to our nation's future.

National Prosperity Comes from Adding Value to Raw Materials

One of the chief causes of the War of Independence was the United Kingdom's policy of requiring its colonies to exchange raw materials for manufactured goods, and these were often not of the best quality. Crow (1943, 24) explains, "The English prohibition against colonial manufacture had been a constant irritant and eventually became one of the causes of [the Revolutionary] war." This reference goes on to add how the newly independent United States went to great lengths to acquire, and "steal" is probably the more accurate word, manufacturing technology from its former master. Pennsylvania offered a 100-pound reward[1] in 1789 to anybody who could invent a power carding machine, and the legislators did not care about the manner in which it was "invented." Crow (1943, 30) adds that one American sawed a British textile machine into flat sections to ship them to the United States while packaged as plate glass. While it is doubtful that the sections could be reassembled into a working machine, they were apparently reassembled into a three-dimensional model for the production of more like it. An apprentice cloth maker named Samuel Slater, whom President Andrew Jackson later called the Father of the American Industrial Revolution, meanwhile memorized a textile machine's design and emigrated to the United States while disguised as a farm hand. The disguise might have been necessary because of British laws against the exportation of manufacturing technology. Emerson (1924, 99–100) makes very clear the advantage of exporting finished goods rather than raw materials and commodities (emphasis is mine).

> The table on page 101, from figures in the June, 1910, report of the United States Bureau of Statistics, shows that one-half our imports consist either of articles of luxury, as silks, wines, diamonds, or of products that do not deplete natural resources, as rubber, sugar, chemicals, or *manufactures of which the value is mainly due to highly skilled labor and delicate machinery,* as cotton and linen lace, works of art and skill; and *that our exports consist largely of prime raw materials, which deplete our natural resources, which are produced in vast quantities by unskilled labor aided by big and rough machines.*
>
> The exported materials, oils, metals, coals, can never be replaced; the exported lumber cannot be regrown in centuries. *The imported silk, sugar, coffee, wool, tobacco and wines consist of brain skill, hand skill, sunlight, air and water; the chemicals are often high priced by-products which we waste; china, glass and laces are immensely valuable compared to the materials which make them, are therefore brain and hand products.* Of the ten leading imported products, diamonds alone are lasting; all the others are fleeting luxuries, eaten up, drunk up, smoked up, worn out before the year rolls around.
>
> *Germany's governmental policy is to encourage the exports of brain, labor, sunshine, air and water*; there is nothing in sugar, in alcohol, but carbon, gathered from the air, but hydrogen and oxygen gathered from the rain water, transformed by the sun into beet plants, grown in fields, tilled and weeded by hand, the beet pulp being transformed by other hands and skilled knowledge into sugar and alcohol. Denmark and Holland export butter which takes nothing from the soil. The French import Asiatic silk, weave it at Lyons, and export the finished product. They export wine, by analysis 87 per cent water, 10 per cent alcohol and 0.04 per cent aroma and bouquet. *Water and alcohol take nothing from the soil, but the aroma makes the wine worth from ten dollars a pound down.*

It is to be noted that ten dollars per pound, or roughly ten dollars per pint, was in the money of the early twentieth century. This was from a process that used free sunlight to turn equally free air and water into fruits, which fermentation could then turn into wine.

Be at the Top of the Food Chain

All manufacturing bills of materials (BOMs) begin with inexpensive raw materials and commodities, combine these to make subcomponents and parts, and assemble the latter to create high-value finished products. Emerson (1924, 98) describes, for example, how the Swiss "imported raw materials from $20 a ton up, and they exported them again as watches worth from $32,000 to $16,000,000 a ton, the difference between import value being Swiss brains and handicraft." We can consider similarly today the difference between the price of a pound of semiconductor-grade silicon and of a pound of semiconductor devices ready for installation in electronic products, computers, and vehicles. Table 2.1 shows the progression in value from raw materials to finished goods.

Figure 2.1 presents a very simplified BOM for power tools, which sell for a lot more per pound than the raw materials and subassemblies that go into them. The idea is to buy the raw materials and make and sell the finished goods.

Buxbaum (2018) reports, however, that the United States has been doing the opposite by exporting scrap iron and steel—commodities at close to the bottom of any bill of materials—to the People's Republic of China (PRC) to feed PRC steel mills instead of American ones. The PRC is now curtailing these imports because it has enough scrap of its own, but the bottom line is that

Table 2.1 Addition of Value to Raw Materials

Raw Material	*Commodity*	*Intermediate*	*Finished Product*
Sand	Semiconductor-grade silicon	Semiconductor devices	Electronic devices including computers, smartphones, and automotive components
Timber	Boards, wall studs, and so on		Buildings
Iron ore	Steel	Steel parts	Steel products
Cotton	Cotton cloth		Cotton garments

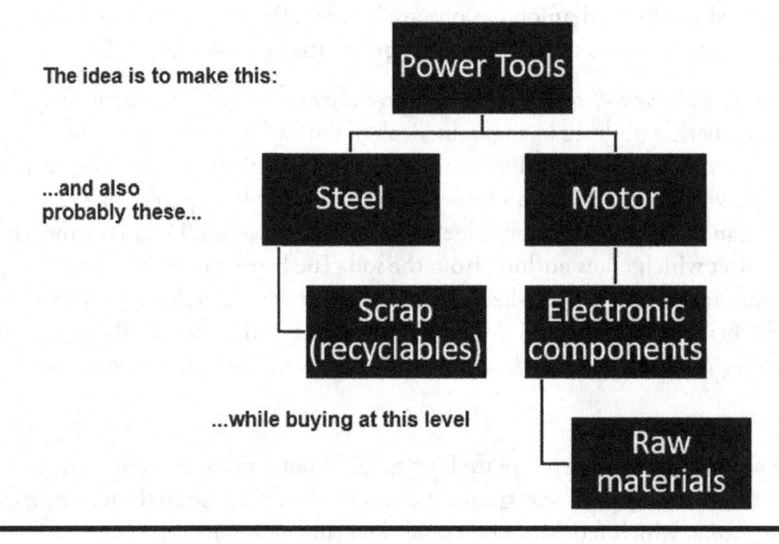

Figure 2.1 Simplified BOM for Power Tools

the United States has (again) been exchanging raw materials for finished goods. The fact that we are exporting scrap metal to the PRC and importing many of the hand and power tools we see in home improvement stores underscores the problem.

We should similarly buy timber, which is readily available in North America, make it into lumber—the only constraint on that involves the capacity of the sawmills—and make the lumber into furniture and other wood products. Buxbaum (2020) adds that the United States instead exports 25 percent of its hardwood to the PRC. I looked at some wooden chairs for sale in the United States and came across a model that sells for more than $130 a pair; it is made in Malaysia. Another set of two sells for more than $170, and it is also made in Malaysia. Yet another pair, this set for more than $190, is made in Vietnam. I am not sure what value is added to the chairs by shipping wood across the Pacific Ocean to Southeast Asia to be made into chairs and then shipping the chairs back to the United States. Is there something particularly exotic about "Made in Malaysia" or "Made in Vietnam" that makes the product more valuable than one "Made in the USA"? Why do American consumers tolerate prices suitable for American-made products on the products of cheap labor?

Ford (1926, 115) wrote almost exactly the same thing about the transportation of wheat 500 miles within the United States to be made into flour and the flour back across the same 500 miles to be made into bread. He said the cost of transportation would have to be carried by the price of bread, and people would consume less. He questioned elsewhere the wisdom of transporting cattle hundreds of miles to meat-packing factories and then sending the meat back. The lesson is even more applicable to transportation across the Pacific Ocean.

The good news, at least for the United States, is that the demand for domestic building construction is now so high that North American sawmills can't keep up with it (Nicholson, 2021). This means that high-wage American trade workers such as builders and carpenters, as opposed to low-wage offshore workers, are adding value to the lumber by converting it into houses and other structures.

The problem is not limited to the United States either. Morello (2020) reports that Australia is exporting timber, a raw material at the very bottom of the industrial food chain, to the PRC. The article includes pictures of logs, as opposed to boards from sawmills, that are destined for export as opposed to domestic use and adds that there are in fact Australian sawmills that could use the logs.

Morello also quotes Roundwood Solutions managing director Steve Telford: "Our forefathers planted the trees; they were planted with a plan to create jobs into the future. It wasn't about growing wood for Asia." This reinforces Emerson's (1924, 94) observation,

> we childishly squander our national resources in exchange for perishable luxuries supplied us by older and wiser men, corporations, and nations, who, not having gifts and prodigal equipment, still use their brains and hands men who trade us sunshine, water, and air for our mined wealth, for our soil's fertility.

Australia therefore seems to be squandering its very patrimony, namely trees that were planted by people who did not expect to live long enough to use them, to buy manufactured goods instead of creating high-wage jobs for Australians. *Any photo of a truck, rail car, or container ship carrying timber from North America or Australia to another country should therefore come across as a danger sign*, unless there is a genuine surplus we cannot use in our own sawmills.

Decision-makers may argue that they are still "making" the top-level product after they contract with an offshore factory to make it for them but, the instant that offshore factory gains control of the means of production, it no longer needs the original business for much of anything except perhaps its brand name and marketing channels. The Roman Empire was similarly invincible as long as it looked to its own citizen-soldiers for security, but then it outsourced its security to mercenaries and allies such as the Visigoths. The latter soon realized that when all you have in the world is the sword your father left to you and his father left to him, and your Roman paymaster has a huge estate and hundreds of slaves to wait on him but no sword, the solution to this unequal distribution of wealth is obvious. When you have the means of production and your paymaster has only money, of which Spain and Portugal had an abundance in the sixteenth century, the solution is similarly obvious. England and Holland applied it to end up in control of the factories, the money, and even the vineyards of Oporto. Portugal had to literally sell part of itself when its imported treasure no longer sufficed to buy what it needed from England.

Raw materials such as cotton and synthetic fibers can similarly be made into fabric which, while a commodity, is still more valuable than the raw material. A clothing factory can then use fabric to make wearable products. This is what the United States ought to be doing instead of importing clothing with expensive designer labels from other countries. One upscale retailer offers a designer sneaker for almost $900 that is at least made in Italy, a high-wage country with skilled workers, instead of the PRC. Another sells an $850 Italian-made sneaker whose principal selling point appears to be its logo. Yet another sneaker, this one with a fancy designer label, sells for almost $800 and is labeled only as "Imported." If it was made in Italy, France, or another high-wage country known for the quality of its apparel, the label would say so. This suggests that price tags suitable for upscale Italian-made or French-made goods are being put onto cheap made-in-offshore-sweatshop products. One might question, in fact, whether $800 or $900 is a suitable price, at least in terms of buying function and utility, for even high-quality sneakers from high-wage countries. The last chapter will address the issue of fancy brand names that add cost but not utility to any product.

It is difficult to obtain contemporary figures for the actual labor hours that go into a shoe, but Stern (1939, 16) reported that the manufacture of 2,000 pairs of shoes required 31,020 labor hours in 1850 but, due to automation, 1,870 hours in 1936. That is, even with the technology of 1936 as opposed to that of 2021, the manufacture of a shoe required less than one labor-hour. There is no reason why shoes of made-in-Italy or made-in-France quality cannot be manufactured and sold in the United States for $150 or even less.

"Distressed jeans" with holes in their knees—the kind one probably can't even give to the Salvation Army—meanwhile sell for more than $200 and are labeled only as "Imported." If people really want to wear jeans that look like rags, it's a lot more cost-effective to buy them for less than $20 from warehouse stores.

The takeaway from this last example is that our country has abandoned the virtues of making high-quality and relatively inexpensive (due to efficient mass production) products with high-wage labor in favor of importing expensive low-quality products including literal rags made by cheap offshore labor. Education of American consumers about the importance of value for their money can help drive reshoring, although manufacturers must still deploy the efficiency principles that were developed here more than a century ago and proven repeatedly to work.

Raw Materials Are Fleeting: Manufacturing Endures

This chapter will discuss later how Spain and Portugal discovered what were essentially boom towns in the New World; venues in which they could get almost limitless quantities of gold and silver to exchange for manufactured goods from England and the Netherlands. The British and Dutch ended up with most of the gold and silver, the factories, and the vineyards of Oporto in the bargain. The fate of Bodie, California, exemplifies the fact that raw materials are ephemeral while manufacturing is forever. Bodie was a typical boom town during the California gold rush. The people who gained the most were those who sold equipment, and often of poor quality, to the prospectors who hoped to strike it rich. When the gold ran out, Bodie (Figure 2.2) became a ghost town.

Chile once similarly earned a lot of money from the export of saltpeter, a source of the nitrate salts that are an important ingredient of fertilizers. There are numerous posters and ads such as "Crops thrive on Chilean nitrate." Then Fritz Haber developed a process with which to obtain ammonia, another source of nitrogen fertilizer that can be made into nitric acid via the Ostwald process, from the atmosphere. The Haber Process was developed in Germany in 1913 and then expanded by necessity when the British cut off exports of Chilean saltpeter to Germany during the First World War. Emerson described previously how Germany transformed air, water, and sunlight into beer and France did the same with wine; now Germany could convert the air itself into fertilizers and explosives. Other countries later adopted the Haber Process, which reduced enormously the demand for world-class Chilean saltpeter. The result was that once-prosperous saltpeter mining towns like Humberstone became uninhabited ghost towns (Long, 2015).

Figure 2.2 A Gold Rush Ghost Town. Bodie, CA, in 1972. (Photo by Dick Rowan, U.S. Environmental Protection Agency, Public Domain as Publication of the US Government)

Germany demonstrated more than a hundred years ago that nitrogen, the predominant component of air, can be turned into valuable chemicals. Now scientists are looking for ways to remove carbon dioxide from the atmosphere to help counteract global warming. This is somewhat more difficult because atmospheric carbon dioxide is roughly 0.041 percent rather than 79 percent, but this, along with the capture of far more concentrated carbon dioxide from combustion processes, could become attractive if something can be done with it. Myers (2022) reports that Matteo Cargnello, a chemical engineer at Stanford University, is developing catalysts that can turn carbon dioxide into saleable hydrocarbons like propane and even gasoline.

A caveat here is that one cannot get something for nothing, and one must put back into the carbon dioxide the heat (enthalpy) one extracted from it by burning the propane, kerosene, or gasoline in the first place. If however one has a source of cheap renewable energy such as solar power in the American Southwest, wind power, or hydroelectric power, this could become economically attractive. *Chemical Engineering Progress* (2022) reports that solar power can be used to convert water and carbon dioxide into kerosene, which is useful as aviation fuel. The current efficiency is low, but it is being improved, and, more importantly, the experimental plant proves it can be done. Peplow (2022) adds that many other companies are looking for ways to use carbon dioxide from combustion and even the atmosphere to "upcycle" it into useful commodities with the aid of inexpensive renewable power. This is simply an extension of Emerson's original commentary about the conversion of the air itself into saleable products. The overall lesson is that the wealth from raw materials, and even precious metals, is fleeting and transitory, while that from manufacturing endures.

The Danger of Manufacturing Inferiority

While mass production began in the early twentieth century, and the Industrial Revolution began in the mid-eighteenth century, the importance of the ability to add value to raw materials was apparent by the mid-sixteenth century. Zamoyski (1987, 175) describes how the Polish-Lithuanian Commonwealth, one of the most powerful nations in Europe, exported commodities such as cloth, beer, and rope in exchange for manufactured goods. "It was essentially the same kind of trading pattern that places third-world countries at the mercy of industrialized nations today." The Polish economist Andrzej Glaber recognized this problem in 1543, roughly 200 years before the Industrial Revolution. Mahan (1890) described how, during roughly the same time period, Spain's and Portugal's discovery of treasure in the New World ruined their manufacturing industries by making it easy for them to buy things instead of making them. Further details will be discussed later.

The United States' Response: Lean Manufacturing

The development of what we now call lean manufacturing in the early twentieth century was not just a reaction to competition from cheap foreign labor; it was a reaction to the very real menace of Japanese competition. Japan's nineteenth-century Westernization included the study of Europe's best military establishments, including the Royal Navy and the Prussian Army. Emerson (1924) reports that Japan studied carefully how Prussia which, in the mid-nineteenth century, had fewer people and a weaker economy than its rivals Austria and France, developed new organizational principles with which to offset these weaknesses and create the German Empire. Emerson noted also that, by the time of the Franco-Prussian War, the Prussians' von Dreyse needle gun (Zündnadelgewehr)[2] which, while a revolutionary innovation in 1841, was outclassed in 1870 by the French Chassepot Rifle. The Prussians won nonetheless for the reasons depicted in the Introduction. Prussia had fewer resources than the French so it had to use these resources far more efficiently.

Emerson (1924, 19–20) goes on to describe how the Japanese adapted Prussia's military organization to their manufacturing establishments, and American industrialists realized that they had to do something (emphasis is mine).

> The greatest example of the power of rational organization and efficiency principles is not in the German upbuilding, but in the Japanese actual creation in a single generation of a great world power. In 1867 Japan was still feudal. The merchants' guild and the thieves' guild were classed together, both beneath contempt. Her peasantry was impoverished; her finest men and women, feudal dependents without initiative. When it was still a treason punishable by death, a few of the Samurai left Japan, not for wealth or amusement or conquest of any kind, but to absorb whatever there might be good in the Western civilization and to bring it back for use in their own beloved country. They conspicuously, consistently, and intelligently put von Moltke's organization into effect in upbuilding their fatherland, and also applied all the twelve principles, which they had probably independently recognized and accepted before they began their quest. In thirty years, Japan with her 40,000,000 people was able to vanquish China with her 400,000,000. In another five years, Russia, the colossus of the North, that had shattered Napoleon I—Russia, the dread of Great Britain, of France, of Germany for 90 years went down in defeat. American sympathies were with Japan, *but scarcely was the war over before the industrial organization of Japan, as much*

> *superior in principle to ours as were her army and navy to those of Russia, began to make us cry out in cowardly fear.*
>
> It is not the flesh and blood and brains of the Japanese that make them industrially dangerous; it is not their money, for they are poor, not their equipment, for they have but little, not their material resources, because they are meagre. *They are dangerous as industrial competitors because we are dragging along under a type of organization that makes high efficiency possible and they are not; because we have not even awakened and they have to the fact that principles applied by mediocre men are more powerful for good than the spasmodic floundering of unusually great men.*

Emerson wrote that Japan had applied "the twelve principles," so it is worth summarizing them here. Twelve are probably too many to guide the reshoring of American jobs that should have never been offshored in the first place, and four look like they can in fact be combined into one. They are nonetheless as applicable today as they were in 1911, if not even more so.

Emerson's Twelve Principles

Emerson begins by comparing systemic and endemic inefficiency to the parasite known as the hookworm and contends that the efficiency of workers of militia age was less than 5 percent, use of materials less than 60 percent, and equipment less than 30 percent. Multiplication yields less than 1 percent, and industrialists ranging from Henry Ford to a modern Japanese firm that had to increase its output of disposable gowns a hundred fold to meet the needs of the COVID-19 epidemic proved Emerson right. He then identified twelve principles through which to achieve results of this nature.

1. *Clearly defined ideals.* This means organizations must set appropriate goals and ask the right questions because, if we set the wrong goals and ask the wrong questions, we will never get the right answers. Russian engineers made the mistake of asking their Tsar (assuming they had a choice) where to build a railroad between St. Petersburg and Moscow. The Tsar, who knew that a straight line is the shortest distance between two points but apparently little else, drew a straight line on the map. The result was a cost of $337,000 a mile in the money of the mid-nineteenth century in contrast to Finnish railroads that cost $23,000 a mile because they accounted for geography and terrain. Emerson then added that:
   - Dysfunctional performance incentives, such as those that encouraged foremen to hire unnecessary workers to do favors for their church members and increase their own importance back then and those that encourage offshoring today, can wreak enormous havoc.
   - The Suez Canal was built with "free" labor under the system of robot or corvée, in which people had to perform unpaid labor in lieu of taxes. The canal was completed five years late and 166 percent over budget.
   - Emerson predicted in 1911 that airplanes would render battleships obsolete, and this was three years before the Royal Navy tested the world's first torpedo bomber.
   - Emerson questioned whether we really "won" the Spanish-American War, which required us to become a Far Eastern power instead of solving domestic problems such as low wages (a major issue 120 years later) and unemployment.

- Emerson (1924, 91) added that, had New York's impressive subways been built with 6-foot gage (the standard is 4 feet and 8.5 inches or 1,435 mm) for 12-foot-wide double-deck coaches, they could have carried four times as many passengers for only a modestly higher capital investment.

2. *Common sense.* What Emerson calls "near common sense" is a shortsighted focus on immediate objectives such as lower labor costs, lower purchase costs, and so on while "supernal common sense" is a long-term focus on items of genuine importance. Near common sense was a major driving force behind the offshoring of American manufacturing.
   - This chapter also cites again the issue of dysfunctional performance incentives. As but one example, a foundry that was rated on the tonnage of output was quick enough to make the big and heavy castings for a large engine but never seemed to get around to making the smaller parts that were equally necessary to build the engine but required eight times as much work per ton.
3. *Competent counsel*
4. *Discipline.* This means adherence to natural laws of human and physical behavior, or what Henry Ford called the Constitution of the Universe and Rudyard Kipling depicted as the Gods of the Copybook Headings. They are impartial, inviolate, and self-enforcing; we go against them at our peril.

> Under the best management there are scarcely any rules and there are fewer punishments. There are standard-practice instructions so that every one may know what his part in the game is, there is definite responsibility, there are reliable, immediate and adequate records of everything of importance, there are standardized conditions and standardized operations and there are efficiency rewards.
>
> (Emerson, 1924, 143)

- This is consistent with ancient Chinese principles of governance as depicted by Sawyer (1993, 122–123):

> In antiquity the Worthy Kings made manifest the Virtue of the people and fully [sought out] the goodness of the people. Thus they did not neglect the virtuous nor demean the people in any respect. Rewards were not granted, and punishments were never even tried.

- The basic idea is that everybody knows what he or she is supposed to do, and why it needs to be done that way, and then it happens naturally. "No rules" does not however mean no processes, standards, or work instructions including standard work; everybody also knows he or she must do a job according to the prescribed instructions and also why this is important.
- Emerson (1924, 153–154) elaborates,

> There is at least one large business aggregation in the United States in which a strike is unthinkable because it is a coveted privilege to be admitted to it as a worker, a catastrophe to be cast out, and so high is the morale that the workers themselves make and maintain standards of conduct far

> stricter than any usual employer would dare to enforce, although he may print and post rule after rule.

- Emerson did not name the employer in question, but Ford (Ford and Crowther, 1922) reported that one of his shops in Britain turned against its own union when it called for the workers to walk out.

> We took over a body plant in which were a number of union carpenters. At once the union officers asked to see our executives and arrange terms. We deal only with our own employees and never with outside representatives, so our people refused to see the union officials. Thereupon they called the carpenters out on strike. The carpenters would not strike and were expelled from the union. Then the expelled men brought suit against the union for their share of the benefit fund. I do not know how the litigation turned out, but that was the end of interference by trades union officers with our operations in England.

- The workers knew that Ford was giving them a square deal and therefore had no further use for the union organizers. Only when Ford's successors ceased to give the workers a square deal in the 1930s did the United Autoworkers gain a foothold at the company.

5. *The fair deal.* No business can remain in business without customers, suppliers, employees, and capital, and all the stakeholders—or what the ISO 9001:2015 quality management system standard calls relevant interested parties—must get a square deal from the relationship. Ford (1926, 20) wrote of this, "Buyer and seller must both be wealthier in some way as a result of a transaction, else the balance is broken" to which Bassett (1919, 34–35) added, "The best sellers will go to any length, they will revamp all their methods, to carry through the principle that lasting trade depends on mutual benefit."
   - Emerson (1924, 199) adds the exact reason that workers, even today, earn far less than they should while customers pay too much and profits are lower than they ought to be. "It is unfortunate that the employer shies at the suggestion of a 10 per cent advance and pays scant if any attention to a 50 per cent inefficiency, two-thirds of which is his own fault." Emerson could indeed write casually of 10 percent wage increases because jobs can easily incorporate inefficiencies of 95 percent or even higher, and this kind of waste hides in plain view because people do not recognize it as waste.
6. *Reliable, immediate, adequate, and permanent records.* Modern quality management systems such as those defined by ISO 9001:2015 and its automotive counterpart, IATF 16949:2016, recognize the importance of documented information and quality records.
7. *Dispatching.* This relates to scheduling of work and production management, and world-class production planning is a prerequisite to small-lot and just-in-time production today. Emerson also mentions production boards, and these were in fact used in the first part of the twentieth century long before anybody ever heard of kanban.
8. *Standards and schedules*; 9, *Standardized conditions*; 10, *Standardized operations*; and 11, *Written standard-practice instructions.* These principles have evolved into what we call "standard work" today, which ensures consistent production of high-quality output. Standardized conditions meanwhile include what ISO 9001:2015 calls "environment for operation of processes" because we know that temperature, humidity, and other factors can affect quality or measurements.

- Emerson (1924, 311–312) cites the effect of standardization on the accuracy and rate of fire of the newest American battleships.

> But the horse-trotting, fire-fighting American stop-watch practice is also in the Navy, and it was realized that if these big guns could be fired four times as fast, it would be very nearly the same as having four times as many guns or four times as many Dreadnaughts, and also that if the skill of aim could be increased four-fold, if four shots would reach the target as compared to one in the older practice, one modern Arkansas or Wyoming, with twelve 12-inch guns, firing four times as fast and hitting four times as often, will, for the time being at least, be 16 times as effective.

- He added that the 12-inch guns could fire two rounds a minute (Wikipedia says the 12-inch/45-caliber Mark 5 gun[3] could manage up to three), which compares favorably to the 12-inch/35-caliber gun installed on the USS *Texas* in 1892; this could manage only one round per minute. A 100 percent increase in the gun's efficiency, whether through design improvements, superior training, and superior standards for the loading drill, or both was therefore like having two ships and two crews for the price of one, at least until rival nations caught on.
- A similar issue arose in the designs of the ships themselves. The USS *Maine*, for example, mounted a pair of ten-inch guns apiece in two turrets on sponsons, which allowed all four to fire directly ahead or behind the ship but only in a very narrow broadside arc. The result was that the United States paid for four battleship guns, and the ship had to carry the weight of four guns, but it could use only two except under very specific conditions. Later pre-dreadnought battleships mounted both turrets on the center line, which allowed all four to fire to either side. The dreadnoughts of the early twentieth century, however, again used wing turrets that could fire to only one side of the ship. The drawback is exemplified by a comparison between the German *Helgoland* design and the Austro-Hungarian *Tegetthoff* design (Figure 2.3). It is clear from this figure that while both nations paid for twelve heavy guns per ship, the *Helgoland* class could use only eight at a time (unless surrounded, which happened rarely if ever) while the *Tegetthoff* class could use all twelve. Two of the Austrian ships therefore deployed the

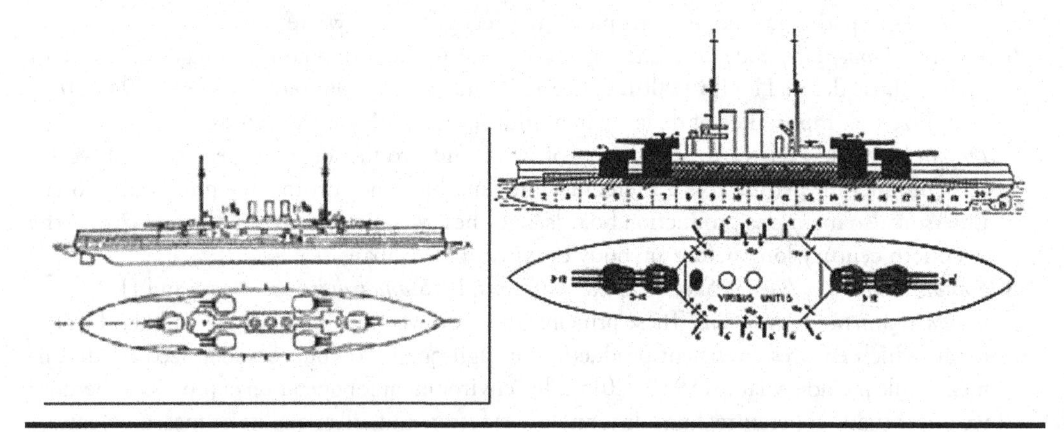

Figure 2.3 Helgoland versus Tegetthoff

same firepower as three of the German ships, so Germany had to pay the capital costs for three ships along with the wages of three crews to get what its ally Austria got from two ships and two crews.

Naval architects finally paid attention to what should have been an obvious consideration, and, as far as I know, no battleship designed after the First World War deployed main battery guns in wing turrets or sponsons. If we carry the same lesson over into industry, it's like having excess and unusable capacity 50 percent greater than what is needed for the job or, alternatively, capacity that cannot be used even if there is a demand for it as there certainly was at Jutland in 1916.

- Had the technology behind the M982 Excalibur guided shell been around at the time, a couple of broadsides from a single battleship could have easily disabled or sunk any conceivable opponent. Near common sense would have probably rejected this idea because of the Excalibur's high cost (Freedburg, 2016); $68,000 for 155 millimeter, although the increased cost for a battleship shell would consist mostly of the cost of the larger shell as opposed to a more expensive guidance package. Supernal common sense would point out that these would actually be a lot cheaper per hit obtained, as anywhere from 95 to 98.5 percent of the unguided shells fired by both sides at Jutland ended up in the water. Supernal common sense would add to this assessment the lives of the sailors and the cost of the ships saved by getting more hits, and more quickly, than the opposing side. The idea is to carry the same principle over into industry where standards and schedules, standardized conditions, and standard operations can create a situation in which one high-wage American worker's productivity is equal to that of any number of cheap offshore workers, and this has in fact been achieved many times. As but one example, a farm worker with a John Deere CP690 cotton harvesting machine can harvest an acre of cotton in six minutes, which makes him or her more than a thousand times as productive as workers who do it by hand—as is apparently still done in the less developed parts of the world. The Case IH Cotton Express is yet another example of how automation makes a handful of high-wage American workers more than equal to any number of cheap offshore laborers.
- Emerson's principles of standards and schedules, standardized conditions, standardized operations, and written standard-practice instructions can probably be condensed for simplicity into a single principle. This is to identify, as Frederick Winslow Taylor (1911) described the matter, the best-known way to do the job (including not just the steps the worker is to follow but also the tools, materials, and suitable environmental conditions for the job) and define this in a written work instruction. When a better way to do the job is identified and proven to work, this becomes the new standard.

12. *Efficiency reward.* Workers are often the driving force behind quality improvements, and the realization of Emerson's principle of discipline, in which everybody does the right thing as a matter of course, requires a square deal for the workers. Ford (Ford and Crowther, 1922) wrote of this,

> But if they see the fruits of hard work in their pay envelope—proof that harder work means higher pay—then also they begin to learn that they are a part of the business, and that its success depends on them and their success depends on it

while Taylor (1911) warned that, if workers did not share in the gains from efficiency improvements, soldiering (marking time, deliberately limiting productivity) would be the natural outcome.

Leaders of the American Response to Japanese Organization

The foremost leaders of the American reaction to the threats of not only cheap offshore labor but also efficient Japanese competition were:

- Harrington Emerson, whose *Twelve Principles of Efficiency* pointed out how enormous waste can hide in plain view
- Henry Ford, the developer of lean mass production
- Frederick Winslow Taylor, the creator of scientific management
- Frank Bunker Gilbreth, the father of motion efficiency, recognized explicitly the application of military motion efficiency principles to civilian enterprises

Their work was sufficiently successful that, by the end of the 1930s, the typical American worker was three times as productive as his or her German counterpart and nine times as productive as a Japanese worker (Ohno, 1988, 3) despite the advantageous position of the latter a mere 30 years ago.

Emerson's observation in 1911 is equally applicable to our situation in 2022. Shortsighted executives whose eyes are solely on the dollar as opposed to the job that creates the dollar have sent jobs offshore to obtain cheap labor even though the United States itself developed the methods and principles with which to overcome competition from cheap labor. Emerson (1924, 93–94) compared "near common sense," which seems to correspond to a focus on the next quarterly report, to "supernal common sense" or a long-term approach.

> A single red copper cent seemed of more worth to the small and terrified soul of a New England statesman than all our splendid country west of the Rocky Mountains, and because he had near common sense, he was willing to sacrifice anything to New England fishing interests; because he was destitute of supernal common sense, he lost to us the empire lying west of the Rockies north of 49 degrees up to 54 degrees 40 minutes, and, no thanks to him, we did not also lose Oregon and Washington.

Emerson was apparently referring to Daniel Webster who, in 1844, opposed the expenditure of $50,000 to establish mail communication with the West Coast. "Mr. President, I will never vote one cent from the public treasury to place the Pacific Coast one inch nearer Boston than it is today." Near common sense, or a short-term focus, said we should not spend $50,000 which was a substantial amount of money in 1844. Supernal common sense, or a long-term focus, would have considered what we could do with the land in question. Near common sense says similarly that it is advantageous to offshore jobs to pay as little as possible for labor; supernal common sense considers factors such as logistics, continuity of operations, and the ability to make jobs sufficiently productive to pay high wages.

Moser (2021) cites the *total cost of ownership* (TCO) which includes hidden costs that are not evident in the product's price. "For example, we recently helped an Illinois printed-circuit-board contract manufacturer save a $60 million order versus a Chinese competitor by showing the customer that the Illinois supplier provided the lower TCO even though it had the higher price." The reference does not go into detail on the hidden costs, but several should be immediately obvious. They include the cost of transportation, carrying costs of inventory in transit (noting that a container ship is essentially a floating warehouse full of inventory), incompatibility with just-in-time production systems, and risks associated with the poor quality for which the PRC is infamous. The total cost of use of, for example, a hazardous chemical can meanwhile far exceed its price even if one accounts for proper disposal of environmental waste. Focus on the price tag is near common sense while focus on TCO is supernal common sense. Chapter 5 will discuss near and supernal common sense further.

Industrial Power Equals Military Power

There is an inextricable connection between the PRC's geopolitical ambitions and the manufacturing capability it has gained at the expense of the United States. Manufacturing has been the backbone of military power ever since the Industrial Revolution, as proven most decisively by the role of American factories in winning the Second World War. A cartoon of the First World War, however, depicted Henry Ford as the Kaiser's most dangerous enemy for the same reason. The Confederacy had better soldiers than the Union, in terms of both marksmanship and motivation, during our Civil War but Northern manufacturing power wore down the South and forced it to surrender. This section will show how the history of warfare has gone through three major phases and also that the most recent should, at least in theory, be the last. The key takeaways are that (1) modern warfare arose from the need to control land for agriculture and (2) the displacement of land by manufacturing as the predominant source of wealth made warfare far more lethal but should also make it obsolete.

1. Hunter-gatherers and nomads often behaved like other animals that will fight, but not to the death, over prime hunting and gathering territory. There is no incentive to escalate to deadly force because the consequence to the loser is simply the need to go elsewhere, with "elsewhere" rarely being far away. Anthropologists have studied primitive tribes that still fight highly ritualized battles whose rules reduce enormously the chance of death or serious injury, much as animals will fight to assert dominance but never escalate the contest to involve the natural weapons they use to kill members of other species.
2. The development of agriculture created both the need and the means to escalate warfare to deadly force. Ancient Greek warfare had very close connections with agriculture because (1) a farmer who was forced off his land would, unlike a nomad or hunter-gatherer, lose his ability to earn a living and (2) farmers had the wealth necessary to afford costly bronze armor. The principal motive for war was almost universally a desire to gain control of somebody else's land. This state of affairs prevailed through medieval times when most of Europe and also the Ottoman Empire demanded military service in exchange for control of land (such as a fief), and even the mid-nineteenth century when the landed aristocracy supplied many countries with military officers. Absolute monarchs meanwhile conducted themselves like the owners of large estates which they sought to enlarge through dynastic marriages and, of course, warfare. People fought for those monarchs not out of patriotism but rather because of feudal obligations or, among the common soldiers, because they were conscripted or could find no other work.
3. The Industrial Revolution enabled countries to equip armies with far more advanced weapons and made industry the backbone of military power. It also, however, displaced land as the principal source of wealth by enabling manufacturers to create almost limitless wealth from raw materials. Henry Ford pointed out almost a hundred years ago that this should have eliminated any rational (i.e., in the aggressor's best interests) motive for war. Events throughout the world show that a good part of the world is not rational—that is, there are people who still believe that military aggression is a good way to get what they want—but manufacturing provides a realistic means of removing the basic root cause of war for the world's rational nations.

Pre-Agricultural Warfare

The connection between industrial and military power is relatively new in human history because, prior to 500 or 600 years ago, a soldier could use only what he and his horse (if applicable) could carry. This meant that typical cottage industries such as blacksmiths, bowyers (trade workers who made bows), and fletchers (people who made arrows) could make everything the soldier needed. The result was that many societies required the soldier to provide his own weapons and, if applicable, a horse. This arrangement also created a close connection between the ownership of land, which was then the principal source of wealth, and military service.

The ownership of land was, in fact, instrumental in the evolution of modern warfare in which soldiers fight to kill one another instead of dominating or scaring off their opponents. Grossman (1996, 6) writes that animals will posture by roaring at one another, threatening one another, and trying to make themselves appear larger than they really are. If posturing will not get another member of the same species to move off, then a fight occurs, but never with the natural weapons with which the animals in question kill members of other species. The risks of doing so far exceed any potential benefits because, if the "winner" of a fight to the death goes away with serious injuries, he or she is likely to be picked off by a stronger predator or else starve due to the inability to forage for food. Piranhas slap one another with their tails, and rattlesnakes wrestle, without using their deadly teeth. The sole consequence to the loser is the need to hunt or forage elsewhere, and humans may have behaved similarly in prehistoric times.

Nomads and herdsmen suffer only displacement from the best land as a result of defeat and therefore have little incentive to fight to the death. Keegan (1993, 29) describes how the pastoral Zulus fought over grazing territory for their cattle, but usually at long range. If a warrior actually killed an opponent, he had to stop fighting and undergo ritual purification lest the opponent's spirit takes revenge on him and his family. "As is typical with primitive people living in underpopulated country, the result was not slaughter but displacement." Only later did Shaka introduce the short spear whose role as a stabbing weapon was similar to that of the Roman gladius and a military organization whose doctrine was similar to that of the ancient Greeks and Romans, namely to kill the other side at close quarters. This enabled the Zulus to defeat opposing tribes who could not even begin to relate to a form of warfare in which an enemy would try to run them through face to face instead of posturing at long range.

Keegan also wrote that the Maring of New Guinea often fought "battles" at extreme archery range to limit the potential damage, and the warriors' wives would collect the enemy arrows so their own side could shoot them back. This is reminiscent of an animated cartoon of a Hatfield-McCoy-type feud in which one side would fire across a valley at the other, and the bullet was too spent to be dangerous by the time it arrived. The other side would then fire the same bullet back, and I recall that the entire feud was thus fought with a single bullet and no fatalities. The Maring could similarly demonstrate their courage in the face of genuine but very limited danger, and if somebody was actually hurt badly or killed, the fight ended for the day. Grossman (1996, 12) adds that the natives of New Guinea removed the feathers from their arrows, which made them virtually ineffective, before waging "war" on one another.

Native Americans in the Plains region of North America and also the Southwest also did not seem to fight pitched battles even though they often used lethal weapons. These Natives, unlike the Aztecs to their South and the Iroquois in the Northeast, relied far less on agriculture and therefore the need to control land. The Sioux, for example, were well known for their ability to convert buffalo into most of what they needed including tents and clothing as well as food. While these Native Americans were clearly willing to escalate to deadly force, as the Seventh Cavalry learned the hard way at Little Big Horn, their warfare seemed to consist more of raids and skirmishes than pitched battles. Hanson (2001, 9) writes, "The most gallant Apaches—murderously brave in raiding and skirmishing on the Great Plains—would have gone home after the first hour of Gettysburg."

Native Americans in the Plains region regarded it as a very high achievement to "count coup," or touch an armed enemy with a harmless stick known as a coup stick, and get away unharmed (Linderman, 2002, 31 offers more details). The entire process looks very much like a way to assert dominance over an opponent without causing him physical harm. All this suggests that whatever passed for warfare prior to the invention of agriculture was mostly nonlethal in nature. The Old Testament says the first murder happened when Cain, a farmer, killed his brother Abel, a herdsman—perhaps not over whose sacrifice was more pleasing to God, but rather because Abel let his animals graze on Cain's land.

People whose livelihoods do not depend on the possession of land, and can carry their possessions on their backs or in horse-drawn carts, can therefore run away from an invader without suffering more than inconvenience. Another way of saying this is that, if a hunter-gatherer loses his or her job of hunting and gathering on prime territory, identical jobs are available nearby in perhaps slightly less desirable, but not particularly distant, locations. A farmer will, in contrast, lose everything he has if he does not stand and fight. This was why ownership of land was once a prerequisite for full citizenship, including the right to vote, in many societies. There was a time in the United States during which only landowners could vote. The landowner had a vital stake in his society—women would not gain the right to vote until much later—while the tenant and migrant did not. The next section will contend that modern warfare, which began thousands of years ago with swords and spears but now carries the risk of use of nuclear weapons, arose from the need to control land as the predominant source of wealth.

Modern Warfare as a Product of Agriculture

Greek hoplites were usually landowners with enough wealth to buy the costly bronze armor that allowed them to stand in the line of battle against their heavily armed counterparts. Recall that, in Homer's *Iliad*, everybody wanted to strip fallen opponents of their expensive bronze armor or prevent that of their friends from falling into the hands of the enemy. Ulysses' goal in Homer's *Odyssey* was to return home to his farm, i.e., the King of Ithaca was nothing more than a prosperous farmer. Ulysses had in fact tried previously to avoid going to the war by feigning insanity, as manifested by plowing a field with salt or alternatively plowing the seashore. Another Greek, Palamedes, exposed this ruse by putting Ulysses's son Telemachus into the path of the plow. Ulysses turned aside immediately, which showed he was not insane so he had to go to Troy after all.

Hanson (1989, 29) explains the Greek hoplite reform,

> By the late seventh century B.C. the security of most of Greek society depended on the arms and armor that each such landholder possessed, hung up above his fireplace, and the courage with which he brought into battle when confronted with an army of invasion encamped on his or his neighbors' farms.

These Greeks fought in heavy armor and used weapons capable of killing at close quarters, and they did not think highly of archers and javelin throwers who fought at a distance. They also usually knew the soldiers next to them as members of their community, and nobody wanted to become a ripsaspis (shield thrower, i.e., somebody who abandoned his shield so he could run away from the enemy more quickly). The light troops who had no real stake in the community were, on the other hand, relatively free to run away if the danger and exertion of pitched battle became too inconvenient for them. The Western preference for decisive pitched battles rather than long drawn-out conflicts also seems to have originated with the Greeks' need to get back to their farms as quickly as possible. They could simply not afford to fight protracted wars that consisted of raids and skirmishes, so they fought extremely violent but relatively short engagements to determine ownership of the land that was the subject of contention.

Medieval feudalism did not relate to feuds but instead to grants of land (fiefs) in exchange for military service. A king or duke would grant a vassal a fief from which the vassal could gain an income and also purchase weapons and a horse. The vassal was required in exchange to present himself for military service during the time of war. King Henry II's Assize of Arms (1181), for example, required his subjects to purchase, at their own expense, weapons and armor according to their standing in society. The holder of one knight's fee (or fief) had to own a suit of chain mail, a helmet, a shield, and a lance, and the same went for wealthy freemen. Less affluent freemen were required to have only a gambeson (a padded jacket), an iron cap, and a lance, and all were required to perform military service at the King's command.

The Ottoman Empire used a similar system in which a Sipahi (Figure 2.4, the name comes from the same Indo-European source as India's Sepoy) was granted the use of some land under

Figure 2.4 Turkish Sipahi. Lorch, Melchior. 1646. Ottoman Sipahi (Public Domain Due to Age)

the Timariot system; *timar* is the word for fief. The timar was not hereditary and could be granted or withdrawn at the Sultan's pleasure. Japanese samurai were similarly granted fiefs in exchange for military service. The king's, Sultan's, or daimyo's control of the land in question ensured the loyalty of his vassals.

This system persisted into seventeenth-century Poland, whose dreaded Winged Hussars were all members of the szlachta (gentry) who were expected to show up for military service with their own horses, armor, sabers, firearms, and retainers including grooms for their horses. The only government-issued item was the light but expensive kopia, a long lance that could outreach infantry pikes. The elected King of Poland, however, had little power because the Sejm (Parliament) decided everything.

The Rise of the Military-Industrial Complex

Agriculture was, as shown above, the principal source of wealth for most of human history and the apparent driving force behind the Western way of warfare. The land-owning aristocracy therefore usually provided an army's officers, and this persisted even into the nineteenth century as depicted by George MacDonald Fraser (1969, 61) in *Flashman*. Lord Cardigan expels the antihero Harry Flashman, the school bully from Thomas Hughes's *Tom Brown's School Days*, from his regiment because of Flashman's shotgun marriage to the daughter of a wealthy factory owner. Flashman protests that his new wife is respectable to which Cardigan, who pronounces l like w, answers, "Could I answer, sir, if I were asked, 'Who is Mr. Fwashman's wife?' 'Oh, her father is a Gwasgo [Glasgow] weaver, don't you know?'" This is despite the fact that Elspeth Flashman's father was probably wealthier than many of the land-owning officers, or the sons and heirs of landowners, in the same regiment. The fact that he owned a factory instead of land meant that he, his daughter, and his son-in-law Flashman would simply "not do" as Cardigan put it.

Some systems including the United Kingdom allowed aristocrats to purchase commissions as the fictional Flashman did, and these officers often looked down on those of equal rank among engineers and artillerymen; that is, officers whose positions depended on technical competence rather than the fortune of their birth. It was, in fact, not possible to purchase commissions in artillery or engineering regiments where the officers actually had to know what they were doing, but only in infantry and cavalry regiments. The Royal Navy did not allow the purchase of commissions either, and again probably because substantial technical competence was required for the jobs. The purchase system, of course, opened the door to people who were not really competent (although many of these were smart enough to delegate actual command to experienced subordinates) but also ensured that the people who controlled the army had a stake in their society.

The evolution of weapons and warfare meanwhile created a situation in which soldiers used more than they could carry, including artillery, while sailors needed warships that were capable of sinking opposing warships. Industry began to play a growing role during the Renaissance and in close connection with military establishments. The Venetian Arsenal (Arsenale di Venezia) was constructed during the Middle Ages and then improved to resemble something similar to a moving assembly line for ships. While blacksmiths could make swords and muskets, larger establishments were necessary for the production of cannons whose manufacture also required quality controls. A defective cannon could easily burst and kill its own crew while rendering itself useless in battle. Juran (1995, 390) adds that Tsar Peter I recognized the need for high-quality weapons and decreed punishments that ranged from flogging and banishment to loss of vodka rations for those who supplied his army with substandard muskets. Peter also founded the Tula Arsenal in 1712 to make weapons for the Russian Army.

It is in fact surprising that the introduction of cannons did not result in the invention of interchangeable parts in the fifteenth or sixteenth century. Cannonballs had to be interchangeable between cannons, which required standards for both the guns and the ammunition. Roser (2015) reports that "go" gages were in fact used to ensure that cannonballs were not too big for the guns for which they were intended; if the ball would not pass through the gage, it would not fit the cannon either. Lieutenant General Jean-Baptiste Gribeauval added a no-go gage that would reject balls that were too small for a good fit. This book will show later that countless people overlook what are, with the benefit of hindsight, obvious improvements. If we paraphrase Sir Arthur Conan Doyle's Sherlock Holmes, many see but few observe. As but one example, tens if not hundreds of millions of soldiers learned motion-efficient loading drills during the horse and musket era of warfare but only in the early twentieth century did Frank Gilbreth suggest the application of the same principles to civilian industries. The ability to notice opportunities that hide in plain view is one of the ways to make high-wage jobs sufficiently productive to keep them in the United States where they belong.

Lepanto (1571) and the Spanish Armada (1588)

Spain's and Portugal's discovery of treasure in the New World at the beginning of the sixteenth century should have theoretically made those countries the most powerful in Europe, but it ruined them instead by making it easy for them to offshore their manufacturing capabilities. The result was that Spain's last great military triumph was at Lepanto (1571), and even that is diluted by the predominant role of the galleasses from the Venetian Arsenal. Spain and Portugal later became little more than parade grounds for British and French armies, both of which came from highly industrialized countries, during the Napoleonic Wars.

The beginning of the end for Spanish naval superiority began during the reign of King Henry VIII who, despite his questionable personal behavior, was a very capable monarch. England has previously imported artillery, but Henry VIII "onshored" its production. It was also during his reign that England developed gun carriages that enabled them to be reloaded far more quickly than their Spanish counterparts. It was during the reign of Elizabeth I that naval architect Matthew Baker built the first *Dreadnought* (1573, the name was later assigned to Admiral John Fisher's brainchild), a razee or race-built galleon that could "run circles around the clumsier Spanish competition" (Boot, 2006). The result was the disaster at Gravelines (1588) which might have been even worse for the Spaniards had the English warships not run out of ammunition. The very fact that the British ran out of ammunition, while many of the surviving Spanish galleons returned home with most of theirs unfired, meanwhile attests to the overwhelming superiority of the British artillery (UPI, 1988). Mahan (1890) explains what happened (emphasis is mine).

> When to these qualities are added the advantages of Spain's position and well-situated ports, the fact that she was first to occupy large and rich portions of the new worlds and long remained without a competitor, and that for a hundred years after the discovery of America she was the leading State in Europe, she might have been expected to take the foremost place among the sea powers. Exactly the contrary was the result, as all know. Since the battle of Lepanto in 1571, though engaged in many wars, no sea victory of any consequence shines on the pages of Spanish history; and the decay of her commerce sufficiently accounts for the painful and sometimes ludicrous inaptness shown on the decks of her ships of war.

[Spain] herself produced little but wool, fruit, and iron; *her manufactures were naught; her industries suffered; her population steadily decreased. Both she and her colonies depended upon the Dutch for so many of the necessaries of life, that the products of their scanty industries could not suffice to pay for them.* "So that Holland merchants," writes a contemporary, "who carry money to most parts of the world to buy commodities, must out of this single country of Europe carry home money, which they receive in payment of their goods." Thus their eagerly sought emblem of wealth passed quickly from their hands ... The fortunes of Portugal, united to Spain during a most critical period of her history, followed the same downward path: although foremost in the beginning of the race for development by sea, she fell utterly behind. "*The mines of Brazil were the ruin of Portugal, as those of Mexico and Peru had been of Spain; all manufactures fell into insane contempt; ere long the English supplied the Portuguese not only with clothes, but with all merchandise, all commodities, even to salt-fish and grain.* After their gold, the Portuguese abandoned their very soil; *the vineyards of Oporto were finally bought by the English with Brazilian gold, which had only passed through Portugal to be spread throughout England.*" We are assured that in fifty years, five hundred millions of dollars were extracted from "the mines of Brazil, and that at the end of the time Portugal had but twenty-five millions in specie"—a striking example of the difference between real and fictitious wealth.

The English and Dutch were no less desirous of gain than the southern nations. Each in turn has been called "a nation of shopkeepers"; but the jeer, in so far as it is just, is to the credit of their wisdom and uprightness. They were no less bold, no less enterprising, no less patient. Indeed, they were more patient, in that *they sought riches not by the sword but by labor,* which is the reproach meant to be implied by the epithet; for thus they took the longest, instead of what seemed the shortest, road to wealth. But these two peoples, radically of the same race, had other qualities, no less important than those just named, which combined with their surroundings to favor their development by sea. They were by nature business-men, traders, producers, negotiators. Therefore both in their native country and abroad, whether settled in the ports of civilized nations, or of barbarous eastern rulers, or in colonies of their own foundation, *they everywhere strove to draw out all the resources of the land, to develop and increase them.* The quick instinct of the born trader, shopkeeper if you will, sought continually new articles to exchange; and this search, combined with the industrious character evolved through generations of labor, made them necessarily producers. *At home they became great as manufacturers; abroad, where they controlled, the land grew richer continually, products multiplied, and the necessary exchange between home and the settlements called for more ships.* Their shipping therefore increased with these demands of trade, and nations with less aptitude for maritime enterprise, even France herself, great as she has been, called for their products and for the service of their ships. Thus in many ways they advanced to power at sea. This natural tendency and growth were indeed modified and seriously checked at times by the interference of other governments, jealous of a prosperity which their own people could invade only by the aid of artificial support—a support which will be considered under the head of governmental action as affecting sea power.

The tendency to trade, *involving of necessity the production of something to trade with,* is the national characteristic most important to the development of sea power.

Mahan therefore states clearly the connection between manufacturing and military power, as demonstrated in the sixteenth century when most combatants, except for sailors and artillery crews, fought with only what they could carry on their backs or horses. The phrase "they took the longest, instead of what seemed the shortest, road to wealth" meanwhile ties back in with Emerson's "supernal common sense" versus "near common sense." The near common sense of the Spaniards and Portuguese told them to collect treasure from the New World to exchange for manufactured goods. The supernal common sense of the British and the Dutch told them that, if they traded manufactured goods for gold and silver, they would end up not only with their factories but also the gold and silver along with the vineyards of Oporto.

Warfare has become far more material-intensive since 1588, but European wars subsequent to England's defeat of the Spanish Armada were waged between countries with roughly comparable manufacturing establishments. France and England were, for example, both capable of producing enough muskets, artillery pieces, and ammunition to equip their combatants. The United Kingdom enjoyed overwhelming manufacturing superiority during the War of Independence, but it also had to fight its former colonies from the other side of the Atlantic Ocean while being at war with France at the same time. The American Civil War therefore appears to be the next major conflict in which a disparity in manufacturing capability was decisive.

The American Civil War

The disparity between the manufacturing capabilities of the industrialized North and largely agrarian South was among the principal causes of the American Civil War, as well as the decisive factor in the Union victory. The South manufactured little, and its economy depended primarily on the export of cotton to England's Lancashire textile mills in exchange for manufactured goods. The Northern factory owners, and also their workers, agitated for tariffs on British imports that were ruinous to the South's economy (Palmer and Colton, 1971, p. 587). Southern cotton production also relied heavily on slave labor, an institution that most of the world's advanced nations had by then abolished. These factors were among the principal root causes of the war.

Once the war began, however, its outcome was almost a foregone conclusion. Barleen (2018) states that the Union had roughly 12 manufacturing workers for every one in the Confederacy and that the industrial outputs of Pennsylvania and New York were each, by themselves, twice as great as that of the entire Confederacy. Arrington (no date given) adds that, as of 1860, "The North produced 17 times more cotton and woolen textiles than the South, 30 times more leather goods, 20 times more pig iron, and 32 times more firearms." The so-called Lost Cause of the Confederacy was exactly that the instant the first shots were fired at Fort Sumter. Stillwell (2019) adds that General Sam Houston warned, "If you go to war with the United States, you will never conquer her, as she has the money and the men. If she does not whip you by guns, powder, and steel, she will starve you to death."

The People's Republic of China, which is among the few remaining nations—the Russian Federation is another—that has expressed a willingness to wage wars of aggression is doing everything it can to put itself in the same position depicted by Houston with respect to the United States. This is consistent with the PRC's playbook, Sun Tzu's *Art of War*, to defeat one's enemy before fighting him.

The First World War

While the role of the Ford Motor Company in the Second World War is well known, its enormous productivity was recognized more than 20 years earlier. A cartoon from the *Illustrated London News* depicted Henry Ford as a "fighting pacifist" throwing a limitless amount of munitions and other war materials at the Kaiser and proclaimed "Henry Ford is the Most Powerful Individual Enemy the Kaiser Has."

The Second World War

The outcome of the Second World War was essentially a foregone conclusion after the United States was compelled to enter it at the end of 1941. Not only did Imperial Japan attack Pearl Harbor on December 7, Nazi Germany declared war on the United States four days later and had the misfortune to be within easier reach than Japan. This was why the war in Europe ended roughly four months before the war in the Pacific.

Taiichi Ohno (1988, 3) noted in 1937 that one American worker was as productive as three Germans, and one German was three times as productive as one Japanese. This had nothing to do with a lack of skill or craftsmanship among the latter and everything to do with the superior industrial organization of the former. Many American jobs were designed, in fact, to not require much skill from the people who did them. Our population was also greater than that of either Axis partner so we had more workers in the bargain.

Hanson (2020) stated that, by 1944, *the United States' gross domestic product exceeded that of all the war's other major belligerents, i.e., Axis and Allies put together.* We out-produced not only Germany, Italy, and Japan but also the United Kingdom and the Soviet Union. We began the war with seven large aircraft carriers, and despite the loss of four that fought in the Coral Sea (*Lexington*), Midway (*Yorktown*), Santa Cruz Islands (*Hornet*), and the Solomon Islands (*Wasp*), we had 27 by the time the war ended. Hanson adds that *the tonnage of the US Navy exceeded that of all the world's other fleets put together* (including the Royal Navy) by the time the war ended.

The Willow Run bomber plant (Figure 2.5), which Ford's production chief Charles Sorensen purportedly designed while he stayed in a nearby hotel, could produce 18 or more B-24 Liberator bombers in a single day. Sorensen (1956, 272) wrote explicitly "The seeds of [Allied] victory in 1945 were sown in 1908 in the Piquette Avenue plant of Ford Motor Company when we experimented with a moving assembly line" and added that Willow Run could deliver one bomber every hour. The disparity between "18 a day" and "1 per hour" is probably due to the fact that the plant had two 9-hour shifts instead of three 8-hour ones.

Manufacturing an End to War

While manufacturing is the backbone of military power, it also offers a way to promote international and domestic peace by eliminating the root causes of war and crime, respectively. George S. Patton Jr. wrote "The End of War" in 1917, which predicted accurately that the First World War was not the purported war to end all wars.

> They will disband their armies
> When this great strife is won
> And trust again to pacifists
> To guard for them their home.
>
> They will return to futileness
> As quickly as before
> Though Truth and History vainly shout,
> "There is no end to War."

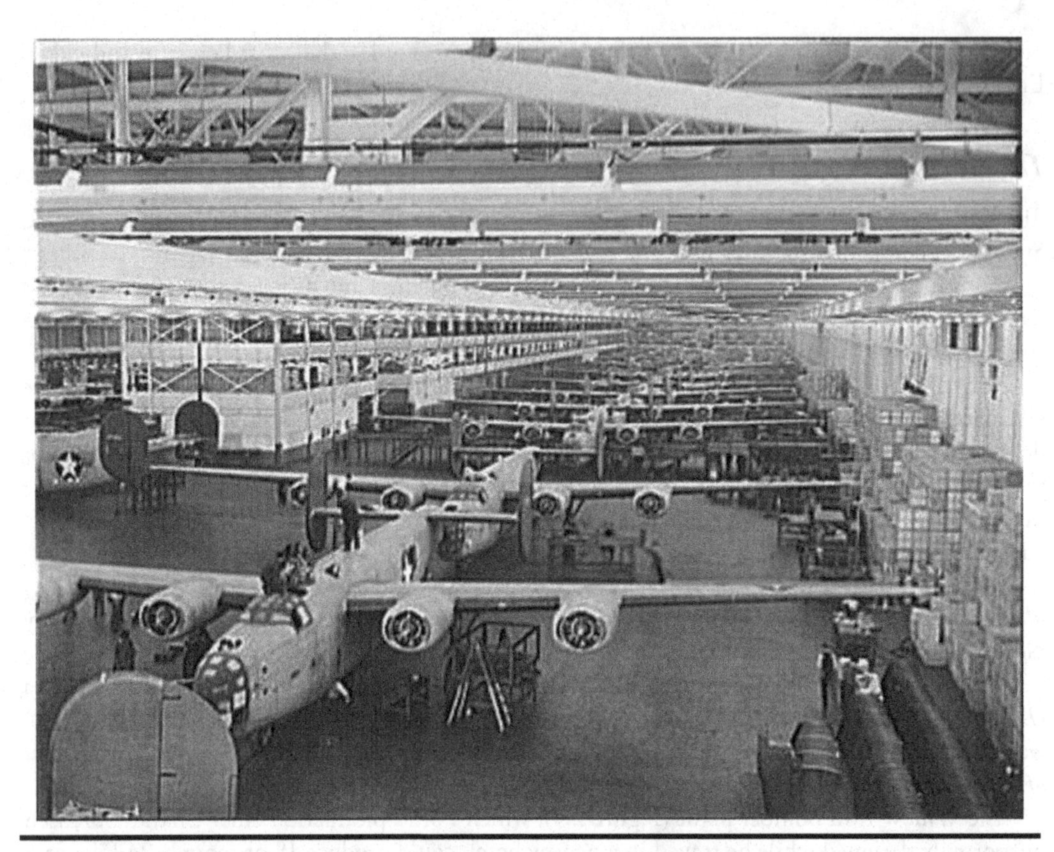

Figure 2.5 Willow Run Bomber Plant, 1942 or 1943 (Public Domain Work of an Employee of the United States Farm Security Administration or Office of War Information Domestic Photographic Unit)

The truth is however that industrialization made war largely obsolete, other than in self-defense, more than a hundred years ago. This coincided, and not by accident, with the enormous productivity gains realized by the Ford Motor Company and other industries throughout the United States and also abroad. Ford (1926, 256) wrote (emphasis is mine),

> The urge to war, springing as it does from the desire to take the fruits of another's production, will always be present, until the peoples of the world have learned to produce in abundance for themselves—*until it has been proven that it is easier to make than to take.*

He added that universal prosperity would render war obsolete and added that the United States had already shown that this could be achieved. Ford's guidance is in fact consistent with what Dr. Stephen Covey (1991, 61–62) calls an *abundance mentality*, which is the premise that resources can be made essentially limitless. The dysfunctional *scarcity mentality*, which is the root cause of most conflicts and wars, assumes that if one stakeholder gets more out of a situation, then the others must by necessity get less. The scarcity mentality also leads to the dysfunctional belief that if workers get higher wages, then customers must pay higher prices and/or employers must accept lower profits.

Two monkeys and one banana exemplify the scarcity mentality. If one monkey has the banana, the other does not so their options are a conflict in which the stronger takes the food from the weaker, or a compromise in which each takes half. Compromise is a lose–lose outcome whose redeeming feature is that everybody involved accepts the outcome as relatively fair. The abundance mentality says that if one monkey allows the other to stand on his or her shoulders to get at the banana tree, both can have more food than they can eat.

None of this means however that Ford was a starry-eyed pacifist because he made it clear that it was neither safe nor realistic to disarm the world's law-abiding citizens while its robbers and aggressors remained armed, and the conduct of the Russian Federation and People's Republic of China in 2022 underscores this caveat today.

Cooperation Is Natural, and Conflict Is Dysfunctional

The abundance mentality is in fact consistent with natural laws that say cooperation is usually far more profitable to everybody involved than conflict. Lethal violence takes place between animals for exactly two reasons, one of which is the need of carnivores to kill their meals and the other is the need of herbivores to use natural weapons like horns and hooves against carnivores.

Cooperation is otherwise natural, and it even takes place between different species. Fishermen in Laguna, Brazil, want to eat fish called mullets, as do the nearby dolphin pods. The scarcity mentality would cause the fishermen to conclude that the dolphins were eating "their" fish so they should kill the dolphins or drive them away. Dolphins with a scarcity mentality might conclude similarly that "those two-tailed monkeys or whatever are eating our fish" (as they might equate the legs of a swimming human to tails) and that they could have more fish if they could get rid of the "monkeys" by ramming them in the water or tipping over their boats to drown them. Dolphins and humans are however both highly intelligent, and they decided collectively to do something entirely different. There are in fact stories to the effect that this arrangement was initiated by the dolphins.

The water in the region is murky, so the fishermen could not see where to cast their nets to catch the most fish. This is not a problem for dolphins, who use biosonar to find their meals, but the fishes' schooling behavior somehow made it difficult for the dolphins to pick off individuals. The dolphins realized, however, that if they drove the fish into the nets, the fish would panic and could be picked off easily. The result was that the fishermen and the dolphins got far more fish through cooperation then they could have gotten from conflict or even independent non-competitive action (Tennenhouse, 2019).

Wolves cooperate similarly to eat large animals that no individual could kill on his or her own. While the pack leaders, also known as the alphas or the breeding pair, eat first, they know that the other wolves (stakeholders in the business arrangement) also must get shares so the relationship will be useful to them. Corporate CEOs who draw large salaries while they look for ways to pay their workers as little as possible would do well to follow the wolves' example instead. Wolves have been around for millions of years while some badly led corporations have lasted but a few decades if even that long.

All of this exemplifies Dr. Covey' s advice to think win-win rather than win-lose, and the win-lose mentality is the root cause of most if not all military conflicts and labor relations problems. If we look at how war has evolved from private quarrels between absolute monarchs to conflicts between industrialized nations, it should be clear that there should no longer be any need for it but also that the world's civilized nations must remain armed against dangerous aggressors that still think they can gain from it.

War Was Once a Private Affair Between Absolute Monarchs

Recall that control of land for agriculture has always been the principal driving force behind war. Absolute monarchs managed their realms like personal agriculture-based businesses and used violence to engage in hostile takeovers of their neighbors' property. They also used peaceful alternatives such as having members of their families marry the right people, and then waiting for the right people to die so they could inherit their land. When Charles the Bold died in battle in 1477, the Holy Roman Emperor moved quickly to marry his son Maximilian to Charles' daughter Mary to add Burgundy to his domains. When this kind of diplomacy did not work, wars of succession often resulted. In 1701, for example, the French Bourbons and Austrian Hapsburgs both claimed that the recently deceased King of Spain had left his realm to them. This resulted in the War of Spanish Succession.

Wars of this nature were essentially private affairs in which ordinary people had little or no interest. People did not join armies out of any sense of patriotism or nationalism, but rather because no other work was available to them. King Louis XIV of France, in fact, had his cannons inscribed with "ultima ratio regum," the final argument of kings as opposed to nations of citizen stakeholders. Clausewitz (1976, 589) wrote, "A government behaved as though it owned and managed a great estate that it constantly endeavored to enlarge—an effort in which the inhabitants were not expected to show any particular interest … War thus became solely the concern of the government to the extent that governments parted company with their peoples and behaved as if they were themselves the state." Clausewitz added that governments relied on cash rather than patriotism to hire "idle vagabonds" to fight for them.

King Frederick II of Prussia (Luvaas, 1966, 72) added, "our armies for the most part are composed of the dregs of society; sluggards, rakes, debauchees, rioters, undutiful sons, and the like, who have as little attachment to their masters or concern about them as do foreigners." The Duke of Wellington depicted his soldiers as the scum of the earth, although he added that the British Army's discipline usually made fine fellows of them.

In the movie version of C.S. Forester's *Horatio Hornblower* starring Ioan Gruffudd, a spy named Wolfe (played by Lorcan Cranitch) joins the Royal Navy under the pretense of fleeing a pregnant woman's angry father. The movie version of Bernard Cornwell's *Sharpe's Rifles* starring Sean Bean has Rifleman Cooper (played by Michael Mears) explain that a magistrate "invited" him to join the army, as an obvious alternative to prison or transportation. Rifleman Harris (played by Jason Salkey) meanwhile enlisted to stay out of debtor's prison, and Sharpe is himself the product of a workhouse and the streets where he learned to fight dirty and ply various criminal trades. Very few men with remunerative occupations had any desire to serve "King and Country," and well-trained and well-paid British merchant sailors did their best to stay out of the clutches of the Royal Navy's press gangs.

Nationalism is, in contrast, a relatively recent development from the late eighteenth century, although Frederick the Great pointed out that a similar sense of what we now call national identity existed in the Roman Republic. Soldiers fought for Rome rather than for whichever two people happened to be Consuls, but this sense of national identity apparently vanished with the fall of Rome. Only with the French Revolution did the modern concept of war between nations of citizens rather than absolute monarchs again become a reality, as depicted by Clausewitz (1976, 592); "The people became a participant in war; instead of governments and armies as heretofore, the full weight of the nation was thrown into the balance." This coincided roughly with the Industrial Revolution, which should have ideally ended the desirability of war.

War in the Industrial Era

While land was still the principal source of wealth throughout most of the nineteenth century, this changed as industrialization became more widespread. Wars were still winnable in the early and mid-nineteenth century, in terms of the victors being collectively better off for having fought. Examples included Italian and German unification, although the American Civil War was ruinous to the entire nation in terms of both human and material losses. Field Marshal von Moltke wrote after Prussia's triumphs in its wars with Austria and France, "On the other hand, who can deny that every war, even a victorious one, inflicts grievous wounds on all involved? Neither territorial gain nor billions in indemnity can replace the dead nor offset the mourning of families" (Hughes, 1993, 22). People were thus beginning to realize that war was no longer worth its cost even in the latter part of the nineteenth century.

Emerson (1924, 79–80) questioned whether the United States was really better off for having won the Spanish-American War in 1898. We obtained the Philippines and Guam, and therefore naval bases in the Far East, but also the costs associated with maintaining a two-ocean Navy. It is quite likely that the Russo-Japanese War (1904–1905) was the last war in which the victors were collectively better off for having fought it, and even that might have taught Japan the wrong lesson; namely that it could do to the United States in 1941 what it had done to Russia in 1904–1905.

Nobody was better off for having fought in the First World War, which made both Nazism and Communism possible. The victors of the Second World War were not better off for having fought although they would have been much worse off had they not, which is simply a restatement of the Duke of Wellington's observation that the only thing worse than a military victory is a defeat.

A factory can add almost limitless value to raw materials, which makes it far easier to trade a small fraction of the finished goods for the materials in question than it is to wage war to control the raw materials. While coal was and is very common, for example, Henry Ford got coke for his blast furnaces for literally less than nothing. He considered it wasteful to burn coal, including the valuable chemicals it contained, for power so he distilled out and sold the chemicals for far more than the coal cost him. Only the use of petroleum products as fuel, i.e., a commodity not much more valuable than the raw material from which it is made, impedes the application of the same principle to oil.

The real tragedy of the First World War was that all the countries involved were doing quite well economically in 1914. None faced a truly existential threat until they mismanaged the crisis caused by the assassination of Archduke Francis Ferdinand and got into a conflict that brought down the governments of four of the major participants (Germany, Austria, Turkey, and Russia). There was meanwhile nothing wrong with Germany's economy and standard of living in 1939 that could be improved by starting another war that ruined Germany for years and kept it divided for even longer. It is difficult to see what North Vietnam "won" from the Vietnam War because, had it chosen peaceful relations with South Vietnam instead, the entire region might now be as prosperous as South Korea and Taiwan.

Hamas meanwhile fires rockets at Israel and suffers the inevitable consequences instead of seeking to imitate Israel's economic behavior to deliver a higher standard of living for the people of Gaza. Israel proved what can be done with the region's limited resources, but Hamas chooses instead to wage a self-destructive campaign of violence rather than benchmark Israel's example. The Russian Federation is now actually losing (as of October 2022) the senseless war it started with Ukraine, and with prominent Russians calling for Vladimir Putin's ouster despite his reputation for violence against anybody who crosses him. The state of affairs is in fact very similar to the one that led to the Russian Revolution in 1917.

The fact that industry should have by now made war an obsolete institution does not mean, however, that the United States can accept a situation in which potential aggressors do not understand clearly that an attack on this country will not have a happy ending for them. This is because, as shown repeatedly by the behavior of aggressors around the world, including most recently the Russian Federation in Ukraine, that many people in positions of power think violence is still a good way to get what they want. The industrialized PRC, whose rulers doubtlessly appreciate the wealth that industry can deliver, cannot for whatever reason restrain itself from brandishing its weapons at neighboring countries and even the United States. The Russian Federation has gone even further by threatening to attack the United States, the United Kingdom, and other nations with nuclear weapons despite the obvious consequences of doing so.

The decisive role of manufacturing, and especially American manufacturing, in victory over the Axis must never be forgotten. The People's Republic of China doubtlessly learned the same lesson, i.e., that industrial power and military power go hand in hand, and this is yet another reason for it to take American manufacturing capability by whatever means, whether honest or dishonest, at hand. American political and industrial leaders must do everything possible to ensure that this agenda does not succeed.

Application to National Social Problems

Even though the United States is a wealthy nation, far too many of its people live in poverty, and often in crime-ridden neighborhoods. If manufacturing can remove the principal root causes of international conflict, it can also remove the root causes of poverty and crime. Ford's industries abolished poverty in the communities where they appeared because the people who worked in them earned very high wages and spent some of these wages in the communities in question. This created societies in which, as Ford (1930, 270) pointed out, the typical criminal worked much harder, and for far less compensation, than a blue-collar worker. Rational people, i.e., those whose goal is to act in their own best interests, would prefer to earn money in decent jobs where there are no risks whatsoever of legal consequences and, in the case of robbery, the prospect of forcible self-defense from a prospective victim. The reshoring of American manufacturing capability will make such jobs available on a widespread basis throughout the nation.

Collapse of American Shipbuilding and Maritime Commerce

Recall that the United States built more naval tonnage during the Second World War than all the other belligerents put together, and we also built cargo ships (such as Liberty ships) more rapidly than the Axis could sink them. Baltimore was meanwhile famous for the construction of fast clipper ships for use in warfare and commerce, and "Baltimore Clippers," the name of an ice hockey team, reflects this history. "U.S. merchant fleet sails toward oblivion" (Little, 2001) speaks for itself: "The exodus of ships has decimated the U.S. merchant marine, once the world's dominant fleet of cargo vessels" and adds that this has resulted in an inadequate pool of sailors with the skills necessary to operate the Armed Forces' cargo fleet during wartime. Alfred Thayer Mahan should have told us all we needed to know about this back in 1890.

Little adds that the US merchant fleet was the largest in the world in 1950, with 3,492 active ships. The number was closer to 220 in 2001, and The United Nations Conference on Trade and Development (UNCTAD, 2020) reports that, as of 2019, 93 percent of the world's ship production came from the PRC, South Korea, and Japan. The United States is therefore no longer even a significant minor player in this industry.

It is also to be remembered that one of the causes of the War of Independence consisted of the United Kingdom's Navigation Acts, which required cargoes to be carried in "British bottoms," i.e., British holds. Madison (1794) later compared the amount of American shipping tonnage and the number of merchant sailors to those of other countries.

> To illustrate this observation, he [James Madison] referred to the navigation act of Great-Britain, which not being counterbalanced by any similar acts on the part of rival nations, had secured to Great-Britain no less than eleven-twelfths of the shipping and seamen employed in her trade. It is stated, that in 1660, when the British act passed, the foreign tonnage was to the British as one to four: In 1700, less than one to six: In 1725 as one to nineteen: In 1750, as one to twelve: In 1774, nearly the same.
>
> On the subject of navigation, he observed that we were prohibited by the British laws from carrying to Great-Britain the produce of other countries from their ports, or our own produce from the ports of other countries, or the produce of other countries from our own ports; or to send our own produce from our own or other ports in the vessels of other countries. This last restriction was, he observed, felt by the United States at the present moment. It was indeed the practice of Great-Britain sometimes to relax her navigation act so far in time of war, as to permit to neutral vessels a circuitous carriage; but as yet the act was in full force against the use of them for transporting the produce of the United States.
>
> Yet in the entire trade between the United States and the British dominions, her tonnage is to that of the United States as 156,000, employing 9,360 seamen, to 66,000, employing 3,690 seamen. Were a rigid exertion of our right to take place, it would extend our tonnage to 222,000; and leave to G. B. employment for much less than the actual share now enjoyed by the United States. It could not be wished to push matters to this extremity. It shewed, however, the very unequal and unfavorable footing, on which the carrying trade, the great resource of our safety and respectability, was placed by foreign regulations, and the reasonableness of peaceable attempts to meliorate it. We might at least, in availing ourselves of the merit of our exports, contend for such regulations as would reverse the proportion, and give the United States the 156,000 tonnage and 9,360 seamen, instead of the 66,000 tonnage and 3,690 seamen.

The infant United States therefore recognized very clearly the importance of maritime trade, a factor cited later by Mahan as a vital element of sea power. The Founding Fathers went to extraordinary lengths to secure the United States' role in international trade, but shortsighted policies squandered this patrimony less than 200 years later. The UNCTAD reference adds that, as of 2019, Asian countries own roughly half of the world's fleet tonnage, the small nation of Greece another 18 percent, and all of North America, i.e., not just the United States but also Canada and Mexico, a paltry 6 percent.

Mahan (1890) also described the relationship between the merchant marine and national prosperity.

> The merchant fleet of Holland alone numbered 10,000 sail, 168,000 seamen, and supported 260,000 inhabitants. She had taken possession of the greater part of the European carrying-trade, and had added thereto, since the peace, all the carriage of merchandise between America and Spain, did the same service for the French ports, and maintained an importation traffic of thirty-six million francs. The north countries, Brandenburg, Denmark, Sweden, Muscovy, Poland, access to which was opened by the Baltic to the Provinces, were for them an inexhaustible market of exchange. They fed it by the produce they sold there, and by purchase of the products of the North,—wheat, timber, copper, hemp, and furs. The total value of merchandise yearly shipped in Dutch bottoms, in all seas, exceeded a thousand million francs. The Dutch had made themselves, to use a contemporary phrase, the wagoners of all seas.
>
> (Mahan cites Lefèvre-Pontalis: Jean de Witt)

Mahan goes on to describe how the Netherlands wrested overseas possessions away from Portugal, the same country that, along with Spain, had ceded its manufacturing power to the Dutch and the British.

> It was through its colonies that the republic had been able thus to develop its sea trade. It had the monopoly of all the products of the East. Produce and spices from Asia were by her brought to Europe of a yearly value of sixteen million francs. The powerful East India Company, founded in 1602, had built up in Asia an empire, with possessions taken from the Portuguese. Mistress in 1650 of the Cape of Good Hope, which guaranteed it a stopping-place for its ships, it reigned as a sovereign in Ceylon, and upon the coasts of Malabar and Coromandel. It had made Batavia its seat of government, and extended its traffic to China and Japan. Meanwhile the West India Company, of more rapid rise, but less durable, had manned eight hundred ships of war and trade. It had used them to seize the remnants of Portuguese power upon the shores of Guinea, as well as in Brazil.

Dutch manufacturing, shipbuilding, and merchant trade made the Netherlands so powerful at sea as to menace the United Kingdom itself as depicted by Rudyard Kipling's "The Dutch in the Medway." "For now De Ruyter's topsails, Off naked Chatham show, We dare not meet him with our fleet—And this the Dutchmen know!" The Netherlands does not however, unlike the United Kingdom, have a body of water in which the troop transports of prospective invaders can be sunk, which left the country vulnerable to invasion by rivals such as France and previously Spain. The power of the British endured while that of the Dutch did not because the Spanish Armada was destroyed in the Channel, the French warships that would have facilitated an invasion were sunk or captured at Trafalgar, and Nazi Germany did not even try. Sir Winston Churchill said of Operation Sea Lion, "We are waiting for the long promised invasion. So are the fishes."

The United States has, despite these lessons of history and through a clearly misplaced set of priorities from its political and business leaders, ceded more than 94 percent of the world's maritime trade to other countries. To put this in perspective, Madison (1794) pointed out how the United States' tonnage exceeded that of every single nation except Great Britain by five or more to one, but now our share of the world's shipping is essentially negligible.

Summary

This chapter has shown that national prosperity and military security are entirely dependent on manufacturing capability. The United States has exhibited dangerous trends that have been associated *universally* with national decline:

1. Offshoring of manufacturing capability in general, and growing dependence on other countries for consumer, industrial, and even military products is extremely dangerous. The next chapter will show that the United States imports components for the latter from the People's Republic of China, and these are often substandard and/or counterfeit by intention.
2. The exchange of raw materials for finished goods is characteristic of the relationship between a colony and its colonial master, or a third-world country and an industrialized one. The idea is to work at the top of the bill of materials by making the high-priced finished product as opposed to the bottom, and the United States has been doing exactly the opposite.
3. Alfred Thayer Mahan described in detail the relationship between manufacturing, merchant shipping, and naval power. The United States still has a powerful Navy but the PRC is catching up, and the US merchant marine has dwindled to the extent that it does not even have enough sailors to meet wartime needs.
4. There is a clear-cut relationship between manufacturing and military power; the second cannot exist without the first. The next chapter will address the behavior of the PRC, which is a dangerous geopolitical rival as well as an infamous human rights abuser and menace to regional and global peace. The wisdom of arming such a nation with the kind of manufacturing base that won the Second World War for the Allies is questionable at best.
5. The ability of factories to generate almost limitless wealth, however, offers real solutions to international and domestic problems, namely the principal root causes of war and poverty, respectively. Rational nations and people, i.e., those that act in their own best interests, that can create wealth lack any real incentive to try to take it from others.

The next chapter will expand on this material by showing that the People's Republic of China is not only an unreliable supply chain partner, it is also a dangerous geopolitical rival to which the United States must not cede any manufacturing capability.

Notes

1. Worth roughly £18,000 today per https://www.in2013dollars.com/uk/inflation/1789
2. The rifle did not use needles (or flechettes) as projectiles, but rather a spring-driven needle that pierced the cartridge to strike a cap behind the bullet. The Chassepot could fire more rapidly and also to a much greater range. The Chassepot was in fact so famous that the original version of William Schwenck Gilbert's song about the Modern Major General from the *Pirates of Penzance* included the original line, "When I can tell at sight a Chassepot rifle from a javelin." Only in 1907 was the Chassepot superseded by the German Mauser of 1898 design in the song.
3. The caliber refers to the length of the barrel in comparison to the bore; a 45-caliber 12-inch gun would therefore be 45 feet long.

Chapter 3

The PRC Is a Dangerous Geopolitical Rival

Chapter 2 depicted the enormous risks associated with the loss of manufacturing capability, and the transfer of this capability to a dangerous geopolitical rival like the People's Republic of China (PRC) makes this even worse. The behavior of the government of the PRC during the past decades, and especially during the past couple of years, is reminiscent of that of Germany, Japan, and the Soviet Union during the late 1930s. None of those countries had designs on the territory of the United States or even the United Kingdom, but their geopolitical ambitions in Europe and Asia resulted in a world war nonetheless.

There is similarly no evidence that the PRC intends to attack the United States directly as stated by Lieven (2020): "Rivalry with China should thus be conceptualized by the U.S. foreign and security establishment as a limited competition in particular areas, not a universal and existential struggle between good and evil." It has however made its regional ambitions very clear by threatening Taiwan, Japan, Australia, and the Philippines the same way Imperial Japan menaced its neighbors in the same region more than 80 years ago. Copp and Baldor (2022) state that a report from the Pentagon warns that the PRC "is working to undermine American alliances in the Indo-Pacific and use its growing military to coerce and threaten neighbors." Shih (2020) reports that the PRC conducted a military exercise complete with a landing craft to simulate an invasion of Taiwan. Regalado (2020) adds that Japanese Defense Minister Taro Kono said the PRC is a security threat to Japan and that there were 177 incidents in 90 days during which PRC military aircraft violated Japanese airspace. The PRC was also, as of April 2021, engaging in provocations against the Philippines in the South China Sea (Venzon, 2021).

The BBC (2020) and Batha (2020) add that the PRC is forcing Uighurs and other minorities to perform slave labor by harvesting cotton. Ordonez (2021) adds that the United States banned cotton imports from Xinjiang Province due to credible reports of slavery. Kuo and Wintour (2020) report that the PRC threatened, "China strongly condemns this [the UK's offer to accept as immigrants all Hong Kong Chinese] and reserves the right to take further measures. The British side will bear all the consequences."

DOI: 10.4324/9781003372677-3

Military threats to neighboring countries, and also a province to which the PRC guaranteed autonomy, are not consistent with the behavior expected from civilized nations in the twenty-first century. The civilized world has meanwhile had laws against slavery for more than 150 years. The fact that three million Hong Kong Chinese apparently want to leave the PRC, and that the PRC must threaten the advanced free nations that would like to welcome them, shows the PRC to be no better than East Germany which had to put up a wall to prevent East Germans from becoming West Germans. It is therefore by no means an exaggeration to compare the PRC's behavior today with that of the Axis and the Soviet Union during the late 1930s, and it is very poor judgment to transfer the industrial might of the United States to an aggressor nation and human rights abuser.

If the PRC can build dozens of aircraft carriers and almost limitless quantities of tanks and fighter aircraft, some of which incorporate stolen American technology (Hollings, 2018), the United States will not dare to lift a hand when the invasions of Taiwan and/or Japan actually come. This is consistent with the advice from Sun Tzu's *Art of War* to win before you fight so that when war (or the threat of war) actually comes, you are facing an opponent who is already defeated. The PRC has meanwhile gained the ability to shut down much of our economy without waging war by embargoing vital exports such as medications and raw materials. When you have somebody by his bill of materials, his heart and mind will soon follow.

The PRC's Threats to American Supply Chains

Buncombe (2020) reports in an article in Xinhua, the PRC's state-controlled news organization, that the PRC threatened to cut off exports of medical products to the United States in retaliation for blaming the PRC for the escape of COVID-19. "Then the United States will be caught in the ocean of new coronaviruses ... If China banned exports, the United States will fall into the hell of a new coronavirus pneumonia epidemic." This is an overt threat to cause the deaths of countless American citizens because of an accusation, which may or may not have been justified, that the PRC allowed the disease to get out of control.

Chakraborty (2020) adds that a PRC export embargo would plunge the United States into "the mighty sea of coronavirus" and that the PRC controls no less than 80 percent of our antibiotic supply and 40 percent of our heparin. It is to be remembered that substandard or even counterfeit heparin from the PRC caused patient deaths in the United States in 2008. Ferry (2020) reports that a PRC economist bragged at the National People's Congress, the PRC's parliamentary body, "we are the world's largest exporter of raw materials for vitamins and antibiotics. Should we reduce the exports, the medical systems of some western countries will not run well." The PRC should never get the opportunity to act on this overt threat.

Minter (2013) adds that a company in the PRC is the sole supplier of the propellant for the Hellfire missile. This reference also cites the US-China Economic and Security Review Commission's 2012 report to Congress that states that the use of PRC-made parts "yields active problems with counterfeit and substandard components and raises the potential for the introduction into critical systems of intentionally subverted components." The PRC has both a motive (degradation of US military readiness) and an opportunity to do this. Whittier and Bordelon (2022) add, "The United States has become too dependent on foreign-made parts, materials, and minerals—especially from China ... If China makes good on its threats to invade Taiwan, we may be entirely cut off from these critical supplies."

The PRC has therefore made it abundantly clear that it means to use its control of life-critical supply chains to threaten the lives of American citizens and is well positioned to do so in the event of a conflict over Taiwan or Japan.

The PRC has subsequently ramped up overtly violent rhetoric in 2022 as shown by commentator Hu Xijin's tweet "Our fighter jets should deploy all obstructive tactics. If those are still ineffective, I think it is okay too to shoot down [House Speaker Nancy] Pelosi's plane" (Reuters, July 30, 2022). While this individual is not a member of the PRC's government, it is also very doubtful that he would have made this statement without his government's tacit approval. When a foreign power whose government controls journalistic and other forms of expression allows prominent individuals to advocate the assassination of a member of the United States Congress, and the same government threatens the United States with war (Inskeep, 2022), we need to end our dependence on that hostile foreign power as rapidly as possible.

The PRC's behavior is in fact very reminiscent of that of Imperial Japan in the same region during the 1930s and needs to be taken just as seriously. The Second World War cut off the United States' access to the natural rubber that was necessary to make, among other things, automobile tires. This led perforce to the development of synthetic rubber that was actually superior to natural rubber, and we need to act similarly and proactively with critical minerals and products (such as semiconductor devices and active pharmaceutical intermediates) that are under PRC control or might fall under PRC control.

The PRC also manufactures 75 percent of the world's polysilicon, which goes into solar panels. Blois (2022) elaborates, "In addition to its massive polysilicon capacity, Chinese companies control the subsequent steps in the supply chain: the production of silicon ingot and wafers, solar cells, and final solar panels." The associated risk is clearly unacceptable if the United States really wants to develop significant solar power capability.

Chinese Communists Lobbied Against Legislation to Promote US Chip Manufacture

President Biden (2022), while speaking on IBM's investment in semiconductor manufacture in New York's Hudson Valley, said of the CHIPS ("Creating Helpful Incentives to Produce Semiconductors") Act,

> China is trying to move way ahead of us in manufacturing [semiconductor devices]. It's no wonder, literally, the Chinese Communist Party actively lobbied against the CHIPS and Science Act—that I've been pushing—in the United States Congress. The Communist Party of China was lobbying in the United States Congress against passing this legislation.

Shepardson (2022) wrote previously, "U.S. Commerce Secretary Gina Raimondo said Wednesday the Chinese government opposes an effort in Congress to ramp up U.S. semiconductor manufacturing because it will give the United States more of a competitive punch." I am not sure why anybody in the United States should care what the Chinese Communist Party wants but the fact that the CCP is against this is an outstanding reason to support it and to reshore semiconductor manufacturing as rapidly as possible. Nikkei Asia (2022) reports meanwhile that Japan is seeking to go "China-free" because Japanese companies do not believe the PRC is a reliable trading partner.

Counterfeit and Substandard Products

The purchase of cheap PRC-made parts may seem attractive on paper if the buyer looks at only the purchase price. Their cost can however be enormous if substandard and/or counterfeit parts result in massive product recalls, user fatalities with consequent product liability litigation and loss of trust in the manufacturer, and failure of military equipment. The PRC-based seller of the adulterated pharmaceutical ingredients or counterfeit semiconductor chips is beyond the reach of the American court system but the end manufacturer is not, so the latter ends up holding the bag.

Ericksen and McKinney (2021) state that many US supply chains are overly dependent on the PRC for the semiconductor devices that go into many electronic products and also vehicles. These critical components are often substandard and/or counterfeit like many other PRC products such as purported N95-equivalent respirators that put their users' lives at risk. Customs and Border Protection (2021) reports,

> U.S. Customs and Border Protection officers working at the Houston Seaport intercepted a shipment of counterfeit N95 masks that with an estimated Manufacturer's Suggested Retail Price of almost $350,000.The shipment of counterfeit masks originated in China and were destined to White Plains, N.Y. before CBP officers intercepted it, April 7.

The PRC counterfeiters used the logo of the National Institute of Occupational Safety and Health (NIOSH) on these respirators without the agency's authorization. As of September 2022, NIOSH's newest additions to its list of respirators that use NIOSH's logo without authorization are all made in the PRC.[1] Many of these are available for purchase over the Internet and have even gotten good reviews from numerous buyers.

OSHA (no date given) warns explicitly of these respirators,

> The use of either counterfeit or altered respirators can jeopardize worker health and safety because there is no way of knowing whether these products or parts meet the stringent testing and quality assurance requirements of NIOSH and will provide workers with the expected level of protection. These products are usually priced lower than certified respirators in order to appeal to cost-conscious employers.

The reference adds that some respirators that failed NIOSH tests were relabeled and fraudulently repackaged as NIOSH-approved.

Nash-Hoff (2012) cites the introduction of counterfeit PRC semiconductor devices into US military products, which undermines our military readiness and puts the lives of our service personnel at risk. Senator Carl Levin said of this,

> Our report outlines how this flood of counterfeit parts, overwhelmingly from China, threatens national security, the safety of our troops and American jobs. It underscores China's failure to police the blatant market in counterfeit parts; a failure China should rectify

to which Senator John McCain added,

> Our committee's report makes it abundantly clear that vulnerabilities throughout the defense supply chain allow counterfeit electronic parts to infiltrate critical U.S.

> military systems, risking our security and the lives of the men and women who protect it. As directed by last year's Defense Authorization bill, the Department of Defense and its contractors must attack this problem more aggressively, particularly since counterfeiters are becoming better at shielding their dangerous fakes from detection.

Simpson (2012) adds that counterfeit PRC parts made their ways into the C-130J cargo plane, Special Operations helicopters, and the Navy's Poseidon surveillance plane. There were 1,800 incidents that involved roughly one million such parts.

Levin (2011) elaborated in detail the enormous risks associated with PRC-made components (emphasis is mine).

> Counterfeit parts from China all too often end up in critical defense systems in the United States. China must shut down the counterfeiters that operate with impunity in their country. If China will not act promptly, *then we should treat all electronic parts from China as suspect counterfeits*. That would mean requiring inspections at our ports of all shipments of Chinese electronic parts to ensure that they are legitimate. The cost of these inspections would be borne by shippers, as is the case with other types of border inspections.

IATF 16949:2016 clause 8.7.1.3, Control of Suspect Product[2] says that *product with "unidentified or suspect status" is to be treated as nonconforming product* which means that, in the absence of very strict controls on traceability and custody, electronic parts from the PRC should be viewed with suspicion.

Remember that the traditional purpose of quality inspections is to detect relatively low levels of unintentional poor quality, as opposed to high proportions of intentionally nonconforming items that are then disguised to conceal their nature. Manufacturers sometimes use "red rabbits," or known bad parts, to test detection controls but, as the name implies, the "red rabbits" are easily identifiable to ensure that they are never incorporated into the end product. Counterfeit parts are known bad parts that are disguised intentionally to cause their incorporation into the final product.

Failure Mode Effects Analysis Perspective

If counterfeit parts are being mislabeled to intentionally defraud purchasers, it is only a matter of time before some get through an inspection to be incorporated into a critical subassembly. Levin (2011) adds, in fact,

> The risk is not created by the contractors. The risk stems from the brazen actions of the counterfeiters. Mr. Kamath of Raytheon, another one of our witnesses, told the committee that "what keeps us up at night is the dynamic nature of this threat because *by the time we figured out how to test for these counterfeits, they have figured out how to get around it.*"

If we look at this from a failure mode effects analysis (FMEA) perspective, the Severity rating of a failure mode that involves a defense-related (or automotive, or civilian aircraft) application is probably 10, the worst possible, on a 1 to 10 scale.[3] This scale is ordinal, which means 10 is not twice as bad as 5; it is worse than 5 by orders of magnitude. A Severity of 5 relates to "degradation of a secondary vehicle function" (*AIAG/VDA*, 2019, p. 109) while 10 relates to something that could kill or injure a vehicle's occupants or, in the context of product realization, result in death or injury to a worker. Ratings of 9 and 10 are reserved for failure effects (consequences of failures) that endanger human life and similarly catastrophic failure effects such as a military product not working when it is needed. The latter failure modes get letters written to the families of our service personnel and lose battles. Failure effects that involve fasteners, active pharmaceutical ingredients, and many semiconductor products are likely to have similarly high severity ratings.

A Severity rating of 8 applies to a product that doesn't work ("loss of primary vehicle function") and also loss of continuity of factory operations for more than one production shift. If counterfeit or substandard semiconductor devices make it impossible to start a vehicle, the vehicle is unusable. This can't endanger human life unless the vehicle is an emergency response vehicle, but it is otherwise the worst that can happen short of danger to life and safety. If a batch of parts is unusable because they are found to be nonconforming or counterfeit, the factory will not be able to make anything.

The *AIAG/VDA* (2019) manual's approach relates the Occurrence rating to the prevention controls whose purpose is to prevent nonconformances. An example would consist of parts whose designs make it impossible to assemble them backward. Prevention controls that make the associated failure mode physically impossible earn a rating of 1 (best possible) while the absence of any prevention controls results in a 10 (worst possible). Even the 10 rating does not, however, account for the *intentional* production and sale of nonconforming or counterfeit products.

The Detection rating reflects the ability of detection controls (such as inspections and gages) to prevent the escape of nonconforming products. If the detection control is certain to catch a nonconforming item, it earns a rating of 1 (best possible). If the nonconformance cannot be detected, the rating is 10 (worst possible). The latter would apply to counterfeit products because the supplier makes intentional efforts to conceal the nonconformances from inspections and tests.

The older FMEA approach multiplies the Severity, Occurrence, and Detection ratings to get a risk priority number (RPN) which can range from 1 to 1,000. Counterfeit products whose failure effects can cause death or injury would therefore get RPNs of 1,000. Wheeler (2011) pointed out, however, that the RPN is the product of three ordinal numbers which limits its utility. Consider two failure modes, the first with $S = 10$, $O = 3$, and $D = 3$ and the second with $S = 3$, $O = 3$, and $D = 10$. Both have RPNs of 90, but the first can kill people while the second will, at most, annoy customers even though the nonconformances are more likely to reach customers because they are not detectable.

AIAG/VDA (2019, 116) uses instead an Action Priority matrix whose deliverable is a priority of Low, Medium, or High. An occurrence rating of 1, which means the failure mode cannot be produced, can earn a Low rating even if the severity rating is 10. If however the occurrence and detection ratings are both 10, as they are for counterfeit products, the action priority will be High for any severity rating of 4 or higher. The obvious question is, therefore, why do we buy anything from PRC sources at all? Levin (2011) elaborates on the PRC's willful efforts to defraud American companies with dangerous counterfeit parts:

> Much of the material used to make counterfeit electronic parts is electronic waste, e-waste, shipped from the United States and the rest of the world to China. E-waste [electronic waste] is shipped into Chinese cities like Shantou in Guangdong Province where it is disassembled by hand, sometimes washed in dirty river water, and dried on city sidewalks. Once they have been washed, parts may be sanded down to remove the existing part number and other marks on the part that indicate its quality or performance. In a

> process known as "black topping," the tops of the parts may be recoated to hide sanding marks. *State-of-the-art printing equipment is used to put false markings on the parts showing them to be new or of higher quality, faster speed, or able to withstand more extreme temperatures than those for which they were originally manufactured. When the process is complete, the parts are made to look brand new to the naked eye.* Once they have been through the counterfeiting process, the parts are packaged and shipped to Shenzhen or other cities to be sold in the markets or to be sold on the Internet.
>
> One of our witnesses today has described to the committee, "*whole factories set up in China just for counterfeiting*" and counterfeit electronic parts are sold openly from shops in Chinese markets.

18 US Code § 2320—"Trafficking in counterfeit goods or services" says that anybody who "traffics in goods or services knowing that such good or service is a counterfeit military good or service the use, malfunction, or failure of which is likely to cause serious bodily injury or death" can get twenty years in prison. Nash-Hoff (2012) adds that 70 percent of the counterfeit parts were from the PRC and originated from 650 different companies there. In addition, the parts are often not even traceable even though traceability ought to be a given. It is mandatory in the auto industry; IATF 16949:2016 clause 8.5.2 (f) extends this requirement to "externally provided products with safety/regulatory characteristics." Even if the application is not automotive, common sense says that anything with safety or regulatory characteristics including active pharmaceutical ingredients (APIs) as well as fasteners, semiconductors, and other components should be subject to the same requirements.

The wisdom of importing military components from a hostile nation known for poor quality and counterfeit products is therefore highly questionable, and the first time that a counterfeit or substandard component was found in a US defense product should also have been the last.

Dangerous Pet Toys and Pet Foods

While dangerous pet toys and treats are far less serious than counterfeit defense-related products, they underscore the PRC's disregard for product quality and contempt for the safety of the end users. Miley (2015) cites not only defective pet toys but also tainted pet foods, many of which come from offshore suppliers, as potential sources of civil liability.

Countless pets did in fact die in the United States because the PRC shipped pet food made with melamine, which resulted in kidney failure. "Melamine is an industrial chemical that has no approved use as an ingredient in animal or human food in the United States" (FDA, 2009). This debacle also resulted in costly product recalls and criminal indictments for introducing adulterated pet food into interstate commerce. Petco and PetSmart pulled all PRC-made dog treats from their shelves in 2015 (CBS News, 2015).

Counterfeit Fasteners in Aerospace and Construction

Benjamin Franklin (1758) wrote of the importance of the seemingly humble fastener, and he was far from the first writer to do so.

> For want of a nail the shoe was lost,
> for want of a shoe the horse was lost;

> and for want of a horse the rider was lost;
> being overtaken and slain by the enemy,
> all for want of care about a horse-shoe nail.

"For want of a rivet" resulted in the loss of more than a thousand lives and an expensive ship when the *Titanic* struck an iceberg in 1912. The National Institute of Standards and Technology (NIST, no date given) reports, "The suspected culprit was one of Titanic's smallest components—the 3 million wrought iron rivets used to hold the hull sections together." The rivets contained far too much slag, with the result that they were too brittle, especially in the cold waters of the North Atlantic, and many broke when the ship struck an iceberg.

The prospect of the loss of an airliner full of people should make the prospect of "for want of a fastener" totally unacceptable but Crouch (1989) reports that counterfeit parts appeared in "nuclear plants, commercial airliners, missiles, trucks, buildings, bridges, school buses and even the space shuttle." Nonconforming ball bearings were sold to Boeing for installation in commercial jets while a US Postal Service building collapsed during an earthquake because 20,000 nuts and bolts were substandard. Had the earthquake happened a few hours earlier, the lives of 400 construction workers would have been at risk. The article adds that some people died in private plane crashes that resulted from counterfeit fasteners. This article cites offenders other than the PRC but, unlike counterfeiters in the PRC, most of these were within the reach of the US Department of Justice. The article does however reinforce the takeaway that counterfeit products are absolutely unacceptable in any supply chain. The Department of Energy (2007) adds that counterfeit bolts, bushings, and brackets caused the loss of a Convair 580 turboprop in 1989 with the loss of 55 lives.

Matters have subsequently improved because of growing awareness and attention to this issue, as well as SAE's AS6832, "Counterfeit Materiel, Assuring Acquisition of Authentic and Conforming Fasteners." It is however known that the PRC and other unscrupulous suppliers are willing to sell dangerous counterfeit items and intentionally disguise their nature to get them past the purchasers' quality controls. Longer and more complex supply chains offer more opportunities for this to happen.

Counterfeit PRC Medications Threaten US Supply Chains

The issue of counterfeiting is hardly limited, however, to semiconductor devices and other components that go into military products. It is also a felony punishable by 20 years under 18 US Code § 2320 to sell counterfeit drugs but, as reported by Lewis (2009), the PRC has exported (as but one example) counterfeit antimalarial drugs to Nigeria. "Often stuffed with chalk, flour or pollen, the pills are passed off as genuine medications. The drug counterfeiters are so skilled that even the holograms on the packages are copied and faked." The fact that the PRC allows or even encourages this should disqualify it as a supply source. Traceability is an issue yet again, and while the Food and Drug Administration (FDA) and its counterparts in other countries (including Nigeria, which took countermeasures) can offer patients considerable protection, the safest course of action is to reshore production into domestic supply chains operated by reputable pharmaceutical firms that are directly subject to FDA regulation.

A set of incidents that resulted in patient deaths was meanwhile traced to adulterated or contaminated heparin (a blood thinner) from the PRC. Stupak (2008) reported (emphasis is mine),

> Today we will hear from two companies responsible for introducing the contaminated heparin into the United States. We will also hear from the FDA regarding the

> circumstances that led to the introduction of the contaminated heparin and its action after the outbreak was discovered. Finally, we will hear from family members of victims who died after being treated with heparin.
>
> It is now estimated that China produces over half of heparin's active pharmaceutical ingredients. *Indeed, all of the tainted heparin in this case was manufactured from API [active pharmaceutical ingredients] produced in China.*
>
> Baxter, the final manufacturer of the contaminated heparin, has a complex international supply chain shown on the slide we have up on the screens. *Their supply chain starts in China, where 10 to 12 Chinese workshops make crude heparin. This crude heparin is then either sold to middlemen called brokers or sold directly to two companies that consolidate the product.*

This again brings up the issue of traceability in complex supply chains, and the absence of traceability for an API should be a bright red warning flag.

> These consolidators then sell the crude heparin to Scientific Protein Laboratories. It is an American company with a plant in Changzhou, China. SPL, Scientific Protein Laboratories, also has a plant in Wisconsin that produces heparin API from the crude heparin. This heparin API is then sold to Baxter, another American company, which manufactures finished heparin products at its Cherry Hill, New Jersey, plant.
>
> In November 2007, Children's Hospital in St. Louis, Missouri, began noticing adverse reactions in their dialysis patients. On January 7, 2008, the Missouri Department of Health and Senior Services notified the Centers for Disease Control and Prevention, who, in turn, notified the FDA and Baxter of the cluster of adverse events.
>
> On January seventeenth, almost 3 months later, Baxter, which produced about 50 percent of the heparin used in the United States, initiated an urgent nationwide recall of nine lots of heparin products after there was an increase in adverse reactions patients suffered while being given heparin products.
>
> On February 11th, FDA announced that Baxter had halted manufacture of multi-dose vials of heparin because of serious allergic reactions and low blood pressure in patients. On that same day, FDA announced that approximately 350 adverse events associated with heparin had been reported since the end of 2007, and the FDA classified 40 percent of these events as serious, including four deaths. *Days later, Baxter recalled all of its heparin injection and solution products remaining on the U.S. market.*

The "inexpensive" PRC-made active pharmaceutical ingredient resulted in a costly product recall, so cheap proved costly in the long run. Stupak (2008) continues,

> As of today, there have been 81 deaths and at least 785 severe allergic reactions associated with heparin since January 2007. Sixty-two of these deaths occurred between November of 2007 and February of 2008.
>
> FDA's investigation into the cause of the outbreak revealed that heparin API made by Changzhou SPL contained a contaminant called oversulfated chondroitin sulfate ... Because oversulfated chondroitin sulfate mimics heparin, it was not detected by

> standard tests. *Oversulfated chondroitin sulfate is not an approved drug in the United States, and it should not have been present in heparin.* In samples collected from Changzhou SPL in China, FDA found that this contaminant was present in amounts ranging from 2 to 50 percent of the total content of the API. The contaminant was also found in some of Baxter heparin lots associated with adverse reactions.

If oversulfated chondroitin sulfate cannot be detected by the standard tests, we now have what looks like a 10-severity failure mode with an equally bad detection rating, i.e., the detection controls cannot catch the poor quality. Even worse, intentional adulteration makes the occurrence rating 10 as well, and the risk priority number (RPN) under the old FMEA system is, as it was for the counterfeit electronic parts, 1,000 out of a possible 1,000.

> To date, it is not known whether this contaminant entered the supply chain accidentally or was introduced intentionally. Because oversulfated chondroitin sulfate is not normally found in nature and is produced through chemical modification, *evidence would suggest that this contaminant was intentionally introduced at some stage in the supply chain.*
>
> While FDA must be applauded for its outstanding efforts in responding to this outbreak, it must also be held accountable for one glaring and fatal mistake: in 2004, a series of FDA blunders resulted in an FDA decision to approve Changzhou SPL to sell heparin HBI to Baxter without first the FDA conducting a pre-approval inspection of Changzhou SPL's production plant, as is the FDA's policy. This plant was not registered in China as a drug manufacturer, and Chinese officials had never inspected the plant either.
>
> It was not until February twentieth that the FDA began an inspection of the Changzhou plant. In that inspection, *FDA determined that Changzhou SPL was incapable of providing safe heparin API to the United States.*

Numerous product liability suits were filed against Baxter which, unlike the PRC sources that defrauded the company, is within the reach of the US judicial system (Michon, no date given). Japsen (2011) adds that one of the deaths resulted in a $625,000 judgment against Baxter because of what Baxter's offshore suppliers did. "A mountain of litigation has been leveled against the companies after U.S. regulators determined in 2008 that Baxter's heparin was contaminated with fake ingredients sourced in China."

Substandard Respirators for COVID-19 Protection

Glatter (2020) reported that 70 percent of KN95 respirators, the PRC's counterpart to the N95 respirator used by healthcare workers and others to protect themselves from inhalation of contagious aerosols, did not meet US filtration efficiency standards. Leo (2020) adds that Health Canada issued a recall of PRC-made KN95 respirators that "pose a health and safety risk to end users." This is not to say that others also did not exploit widespread, and justified, fears of COVID-19 to sell counterfeit and substandard respiratory protection but a very large share was traced to the PRC. Note again that the sellers are almost certainly beyond the reach of US and Canadian jurisdiction.

While the PRC's cheap labor might make these items nominally cheaper, these incidents underscore the enormous risks associated with supply chains over which the end manufacturer has little control and whose elements are willing to cut corners at the expense of product safety. This is not to say that domestic suppliers have never cut corners either but, unlike suppliers in the PRC, they are within easy reach of our regulatory agencies as well as our judicial system. The PRC suppliers can, on the other hand, commit what the US Code defines as serious felonies because they are beyond the reach of US law enforcement as well as our civil courts. The American end users they defraud, such as our pharmaceutical and defense firms, similarly have no recourse against them. The lesson seems obvious; friends don't let friends buy parts or ingredients that can affect customer or patient safety from the PRC.

The next section will show, however, that even were it not for the PRC's threats to regional peace, explicit threats to disrupt US and other supply chains, and export of substandard and counterfeit products, it is still bad judgment to offshore critical manufacturing capability even to friendly, reputable, and quality-conscious trading partners.

General Supply Chain Risks

This chapter has shown so far that the PRC is a willfully unreliable supplier that has (1) threatened to intentionally disrupt vital supply chains and (2) knowingly sold US customers, including defense contractors and pharmaceutical companies, substandard, adulterated, and/or counterfeit products. The COVID-19 epidemic along with a steady series of natural disasters and other force majeure ("greater force") incidents, including a ship getting stuck in the Suez Canal, underscores however the risks associated even with reliable and trustworthy suppliers in complex supply chains. This is another strong argument for reshoring to the United States where, while force majeure is still a risk—a snowstorm might close the highway that trucks use to make deliveries or the Susquehanna River might overflow its banks—it is less serious than that associated with suppliers on the other side of an ocean.

IATF 16949:2016 clause 6.1.2.3, Contingency Plans, requires organizations to develop contingency plans for *continuity of supply* in the event of force majeure and with good reason. The leanest and most advanced factory on earth cannot make anything if it cannot get even one item in a complex bill of materials or a vital consumable such as the resins used in the manufacture of semiconductor devices. Norwood (1931, 35–37) describes how the Ford Motor Company had very extensive contingency plans to re-route rail shipments to avoid plant shutdowns in the event of a force majeure situation. This was achieved, of course, without the benefit of today's computerized logistics systems.

Semiconductor and Automotive Supply Chains

Taiwan is, unlike the PRC, a free nation to which Freedom House gave a score of 94 out of 100 for 2021 in contrast to the PRC's abysmal "Not Free" rating of 9. This does not mean however that Taiwanese, Japanese, and other responsible offshore suppliers are not vulnerable to force majeure. Agence France-Presse (2021) reports, for example, how a drought in Taiwan is interfering with the manufacture of semiconductor chips. The manufacture of semiconductor products requires enormous quantities of high-purity water that is free of not only dissolved minerals but also ionic contaminants, and Taiwan has not been getting enough rainfall to meet the need. That is, a shortage of something seemingly ordinary like water can interfere with a complicated supply chain; the phrase "for want of a nail" again comes to mind. *Quality Progress* (2021) reports meanwhile, "Chip Shortage Causes Automakers to Hit Pause," which reinforces the fact that any supply chain interruption can cause billions of dollars in lost economic activity. *Industry Week* (2021) reported meanwhile, "Toyota Announces 40% Worldwide Production Cut Due Next Month" because COVID-19 surges in Malaysia and Vietnam interfered with semiconductor supplies. The opportunity cost (revenues foregone because roughly 400,000 vehicles are not manufactured or sold) probably exceeds ten billion dollars, which leads to the question as to why the necessary devices are not made in Japan and the United States, where Toyota has some factories.

Reuters (2022) meanwhile reported, "Ford to halt production next week at Flat Rock plant on chips shortage." One cannot build Mustangs without semiconductor devices. Wayland (2021) adds, "GM and Ford cutting production at several North American plants due to chip shortage."

Henry Ford used vertical integration, in which he controlled most if not all of his supply chains from raw materials to finished goods, for a reason. While true vertical integration might not be possible or feasible today, there are few if any reasons to not make semiconductor devices domestically, or at least in North America, to ensure continuity of supply.

These are but the most recent incidents of many that have affected semiconductor supply chains. Savitz (2011) reports that, in 2011, an earthquake (force majeure) in Japan interrupted one-quarter of the world's supply of the silicon wafers that are used in semiconductor manufacture. Powell (2011) adds that autoworkers in Ohio suffered a cut in working hours because Honda could not get parts from Thailand because a devastating flood temporarily rendered electronics component maker Rohm & Co. unable to deliver them. APICS (2011) reported that 85 percent of supply chains experienced disruptions in a single year.

This is not to say that floods are not a problem in the United States, and residents of Northeast Pennsylvania are quite familiar with the Susquehanna River's infamous reputation for overflowing its banks. Levees were improved to protect Wilkes-Barre after a good part of it was destroyed by Hurricane Agnes (1972). This did not however protect Bloomsburg where a subsequent flood idled several automotive plants due to a lack of vehicle carpeting (*Automotive News*, 2011). This was not however allowed to happen more than once, as automakers insisted that action be taken to prevent a recurrence, and it is reasonable to believe they have far more influence over domestic suppliers than offshore ones. The measures taken ranged from stormwater basins to levees and pumping stations (Lawson, 2017).

COVID-19 has also caused supply chain problems. Krisher (2021) reports that the delta variant of COVID-19 has resulted in shortages of semiconductor devices from Asian suppliers, which leads again to the question as to what benefit there is to making these devices offshore at all. The shortage of chips has led to a shortage of vehicles, with new vehicle prices averaging above $41,000, which many buyers are probably not willing and/or able to pay. The same reference adds that automakers have allocated the scarce chips to their higher-priced vehicles which means lower-priced ones are in short supply.

Other Supply Chain Problems

Smialek and Ngo (2021) report that Catrike has 500 three-wheeled bicycles ready for sale, except they are lacking derailleurs (parts of a bicycle's drive train) that are made in Taiwan. The result is that $2 million in inventory is waiting on parts that cost $30, which brings to mind immediately the story "For Want of a Nail." Complex international supply chains that involve even reliable suppliers such as those in Taiwan, let alone often-unreliable ones in the PRC, increase supply chain risks enormously.

D'Innocenzio (2021) adds meanwhile that the Basic Fun toy company could not ship a third of its Tonka Mighty Dump Trucks from the PRC to the United States due to high prices for shipping containers. The reference adds that the cost of overseas transportation eats up 40 percent of the retail price, and this does not even account for the cost of subsequent domestic transportation. Recall that the total cost of ownership (TCO) or total cost of use of something can far exceed its ostensible price tag when one accounts for costs of this nature. The reference adds further that supply chain problems could impact the 2021 Christmas shopping season, upon which retailers depend for an average of 20 percent of their sales. Bottlenecks have also appeared in the seaports through which all the PRC-made goods must flow, unless the importer wants to pay for air transportation instead. The first question that comes to mind is why the manufacturing jobs were outsourced to the PRC in the first place.

Goodman (2021) discusses meanwhile how seaports are overflowing with inventory because there are not enough truck drivers to move the deliveries. The article discusses the arrival of a Chinese-name container ship, "its decks jammed with containers full of clothing, shoes, electronics and other stuff made in factories in Asia. Towering cranes soon pluck the thousands of boxes off the ship—more cargo that must be stashed somewhere." Is there something special about clothing, shoes, and electronics made in Asia, and probably the PRC, that makes them more valuable than the same goods made in the United States? "Cargo that must be stashed somewhere" is meanwhile inventory, with all the drawbacks of carrying costs and obsolescence risks.

While the prospect of force majeure is not as serious as the PRC's overt threats to cripple supply chains by withholding critical materials and components, and its long record of defrauding customers with dangerous substandard and/or counterfeit parts, it is still a strong argument for reshoring vital consumables and bill of materials components to the greatest extent possible.

Summary

This chapter has discussed the enormous risks of offshoring manufacturing capability to a dangerous geopolitical rival with a well-deserved reputation for the intentional sale of counterfeit and substandard products. It has also shown, however, that even trustworthy offshore suppliers are often subject to force majeure and that supply chain interruptions can occur regardless of their best efforts.

The next chapter will show, however, that even the prospect of saving money on labor even in the absence of the indicated risks could in fact be a dangerous illusion. It has been proven repeatedly that high-wage labor can be made sufficiently productive to make wages almost irrelevant to the cost of the product.

Notes

1. https://www.cdc.gov/niosh/npptl/usernotices/counterfeitResp.html
2. International Automotive Task Force; IATF 16949:2016 is the ISO 9001:2015 quality management system standard with additional requirements that are well worth reading by ISO 9001 users.
3. Failure Mode Effects Analysis is a quality planning process that seeks to identify proactively what might go wrong (the failure mode), the consequences (the failure effect), and the underlying reason (the failure cause or mechanism). The deliverables of this planning process include design and production controls whose purpose is to disable the failure modes, or at least intercept them before they can affect customers or users.

Chapter 4

Cheap Labor Is a Dangerous Illusion

We have seen so far that the offshoring and decline of American manufacturing capability, with much of it going to a dangerous geopolitical rival, is a clear and present danger to our nation's affluence and military security. The next step is to determine why this happened and to address the dysfunctional paradigm that cheap labor enables low prices and high profits. The truth, while counterintuitive, is that *cheap labor is more frequently symptomatic of excessive prices and low profits.* There are two reasons:

1. Almost any job can be made sufficiently productive so that the contest is no longer between high-priced workers and cheap or even unpaid ones but between efficiency against inefficiency and machines against manual labor.
2. Cheap labor gives enormous amounts of waste a perfect place to hide, and the low-wage workers have no incentive to do anything about it. Frederick Winslow Taylor's goal was to select what he called "high-priced men"[1] or workers who would follow to the letter directions from efficiency experts in return for premium wages. He did not expect manual laborers to exercise initiative to improve their jobs, but we now recognize that many if not most improvements come from frontline workers.

Meet the High-Priced Workers

The original context of Taylor's (1911) high-priced workers was his proposition to an uneducated pig iron handler that if the latter was a "high-priced man," he would do exactly as he was told.

> Well, if you are a high-priced man, you will do exactly as this man tells you tomorrow, from morning till night. When he tells you to pick up a pig and walk, you pick it up and you walk, and when he tells you to sit down and rest, you sit down. You do that right straight through the day. And what's more, no back talk.

DOI: 10.4324/9781003372677-4

> Now a high-priced man does just what he's told to do, and no back talk. Do you understand that? When this man tells you to walk, you walk; when he tells you to sit down, you sit down, and you don't talk back at him. Now you come on to work here to-morrow morning and I'll know before night whether you are really a high-priced man or not.

This is probably the source of the perception that Taylor told workers to leave their brains at the factory gate, which is not consistent with today's most advanced practices. Taylor elaborated, however, "This seems to be rather rough talk. And indeed it would be if applied to an educated mechanic, or even an intelligent laborer." He wanted the uneducated laborer to focus on the high wages he would earn by doing exactly as he was told, and, when he paced himself as directed, he was able to load 47 rather than 12.5 tons of pig iron onto a rail car every day. His daily wage meanwhile increased from $1.15 to $1.85 a day, which was a substantial amount in the late nineteenth and early twentieth century. The truth is meanwhile that doing exactly what one is told, e.g., by a work instruction or process, is the foundation of standardization.

High-Priced Workers and Standards

While the concept of doing exactly what one is told seems to be most suitable to unskilled workers, the commercial airplane pilots with whom we entrust our lives do exactly as they are told (by a set of instructions) when they follow pre-flight checklists. The same goes for many if not most jobs in the Armed Forces; everybody follows the prescribed instructions which are the best-known ways to do a job. Knowledge workers such as engineers and technicians also conform to standards when, for example, they follow a process for failure mode effects analysis (FMEA) or corrective and preventive action (CAPA) even though they must also apply judgment, experience, and education to their work.

This does not mean however that standardization makes workers the equivalent of mechanical robots. Standards are the foundation upon which intelligent organizations can build improvements, and there is nothing upon which to build if there is no foundation. *Standard Work for the Shopfloor* (2002, 24) explains, "Standardization is not only adherence to standards but also the continual creation of new and better standards." Ford (1926, 82) added, "Standardization means nothing unless it means standardizing upward … Today's best, which superseded yesterday's, will be superseded by tomorrow's best." Taylor (1911) supported this position and also the proposition that workers should indeed think for themselves.

> It is true that with scientific management the workman is not allowed to use whatever implements and methods he sees fit in the daily practice of his work. Every encouragement, however, should be given him to suggest improvements, both in methods and in implements. And whenever a workman proposes an improvement, it should be the policy of the management to make a careful analysis of the new method, and if necessary conduct a series of experiments to determine accurately the relative merit of the new suggestion and of the old standard. And whenever the new method is found to be markedly superior to the old, it should be adopted as the standard for the whole establishment. The workman should be given the full credit for the improvement, and should be paid a cash premium as a reward for his ingenuity. In this way the true initiative of the workmen is better attained under scientific management than under the old individual plan.

Standardization and standard work therefore do not stifle the workers' initiative; they empower the workers to use them to drive continual improvement.

When the employer offshores jobs to low-wage countries or pays the lowest wages possible, however, the workers have no incentive to make any improvements or do anything other than what they are told. This is why, as long ago as the Biblical story of Gideon, intelligent leaders have selected high-priced people over cheap ones.

High-Priced Soldiers Are Cheaper Than Cheap Soldiers

The prospect of invading somebody else's country with hundreds of thousands of cheap soldiers may sound attractive until one realizes that they will lose their combat effectiveness, and eventually die, unless they can be provided with food and other necessities. The same goes, in fact, for hundreds of thousands of high-quality soldiers if one cannot supply them. Russia used this principle successfully against the Swedes (culminating in the Battle of Poltava), the French, and later the Germans. Longworth (1965, 220) describes how Field Marshal Alexander V. Suvorov, whose record consisted of 63 battlefield victories and no defeats, wrote in his *Science of Victory*, "A trained man is worth three untrained: that's too little—say six—six is too little—say ten to one." Suvorov won 63 major battles, including a few essentially one-sided massacres (e.g., Rymnik in 1789 where the Turks lost 20 casualties for every Russian or Austrian under Suvorov's command) through superior training, motivation, and empowerment. Russian autocrats like Tsar Nicholas II, Joseph Stalin, and most recently Vladimir Putin abandoned Suvorov's principles by sending poorly trained, poorly motivated, underequipped, and incompetently led soldiers to their deaths. Putin's conscripts are, as of October 2022, not faring well against the superior training and morale of the Ukrainians while supply is a major problem for the Russians. There is now a running joke that the once-feared Russian Army is not the second strongest in the world, but rather the second strongest in Ukraine. The key takeaway is that one soldier (or worker) with superior training, motivation, equipment, leadership, and organization can easily be worth ten or even more who lack these things.

Even if the soldiers are unpaid conscripts or self-supported volunteers, they still need supplies. The Battle of Watling Street (60 CE) in which two Roman legions, or roughly 10,000 soldiers, under the command of Gaius Suetonius Paulinus, annihilated 100,000 Britons led by Queen Boudica, illustrates this principle. The story, as told by the Roman historian Tacitus, is that Suetonius set up a well-defended position in a narrow gorge that limited the number of Britons who could actually fight. This meant his far better equipped and trained soldiers would not have to fight more than their own number of Britons, and perhaps less because the tightly packed Roman formations usually outnumbered their enemies at the point of contact, the only point that counts. The Romans fought shoulder to shoulder with short swords (gladii) designed expressly for this purpose while their less sophisticated enemies fought as individuals with larger swords that required far more room to wield effectively. This meant that, in a front of any given width, there were far more Romans than opposing combatants.

The first question that comes to mind is, of course, why Boudica played Suetonius' game by attacking at all. Her best strategy, had it been possible, would have been to encamp in front of the Romans to wait for hunger to force them out so her army could attack them from all sides. She may have known that another Roman legion (II Augusta) was still at large somewhere but even its arrival would have left her with roughly five or six to one numerical superiority. The issue could be quite likely that as her army had brought family members who wanted to watch the battle, her own side would have run out of food long before the Romans. While circumstances did not permit her to bring what Frederick Winslow Taylor would later call "high-priced men" to match those of Suetonius, she could have probably left the family members and also her weakest combatants behind.

The same consideration almost cost the Persians the Battle of Thermopylae (480 BCE) in which King Xerxes invaded Greece with, as claimed by Herodotus, more than a million soldiers. They encountered a strongly defended position with only a few thousand Greek defenders, the most famous of whom were King Leonidas's 300 Spartans. The position was roughly 15 meters wide, or wide enough for perhaps 15–25 soldiers to fight side by side, and the Spartans used a tightly packed formation similar to the one used later by the Romans. This made the latter practically invincible until the Persians discovered a goat path that allowed them to encircle the Spartans. The result was that Xerxes, as portrayed by David Farrar in the 1962 movie, lamented that the Spartans were slaughtering his men like sheep.

Why did Xerxes attack such a well-defended position in the first place? Carey (2019) pointed out that while the defenders had a secure supply of food and water, "Xerxes had a huge army to feed and water (it was high summer), plus camp followers and servants, cavalry mounts, baggage animals, cattle and a large and lavish royal court." Xerxes therefore, like Boudica, had to choose between fighting at a terrible disadvantage, retreating and demoralizing his army, and starving where he stood. *Numerical superiority is therefore worse than useless if its possessor cannot bring it to bear against his or her opponent.* The same principle carries over into industry where the numerical superiority of cheap workers requires more floor space, more utilities, and more overhead to go with them.

Gideon and His 300 "High-Priced Men"

The Biblical story of Gideon (King James Bible, Book of Judges) reports how, in contrast, Gideon got rid of his cheap soldiers until only the best ones remained. God presumably counseled Gideon, "The people that are with thee are too many," and perhaps for reasons other than that given. Gideon's first move was to tell those who were fearful and afraid to return to their homes, and he eventually found himself with 300 "high-priced men" who were more than up to doing the job.

> Then Jerubbaal, who is Gideon, and all the people that were with him, rose up early, and pitched beside the well of Harod: so that the host of the Midianites were on the north side of them, by the hill of Moreh, in the valley.
>
> And the Lord said unto Gideon, The people that are with thee are too many for me to give the Midianites into their hands, lest Israel vaunt themselves against me, saying, Mine own hand hath saved me.
>
> Now therefore go to, proclaim in the ears of the people, saying, Whosoever is fearful and afraid, let him return and depart early from mount Gilead. And there returned of the people twenty and two thousand; and there remained ten thousand.

The story continues with a very complicated night attack in which the 300 soldiers had to divide up into three companies of 100 each and, upon command, unmask lanterns and blow trumpets to cause the Midianites to panic. It is quite likely that had Gideon showed up with 32,000 people, some of whom were afraid of the enemy and lacked self-discipline, something would have gone wrong such as somebody unmasking his lantern at the wrong moment, crying out in panic to alert the enemy, or perhaps mistaking friend for foe in the darkness.

The issue of discipline, in fact, came up when Gideon dismissed the soldiers who knelt to drink from a river while others took the more dignified approach of using their hands to bring the water to their mouths. "And the number of them that lapped, putting their hand to their mouth, were three hundred men: but all the rest of the people bowed down upon their knees to drink water." These 300 were the "high-priced men" upon whom Gideon could rely to do the required job with no mistakes. If we return to the initial point, meanwhile, Gideon had to feed 300 rather than 32,000 soldiers which made the logistics aspect much simpler.

The same principle carries over into industry where, for example, a company that hires 1,000 workers at minimum wage rather than 300 high-wage ones incurs costs such as mandatory employment taxes, safety training, floor space, restroom space, supervision, and so on for the 700 additional people. The workers who know they are getting as little pay as their employer can deliver will meanwhile do only what their supervisors tell them to do and perhaps only when the supervisors are watching. This alone increases the overhead cost of supervision. The same workers are unlikely to care about the quality of their work, or whether waste and/or poor quality are built into their jobs. Why should they? Taylor (1911) added that such workforces even used peer pressure to discourage coworkers from trying to improve the jobs. The following is the exact opposite of what we seek to achieve with lean manufacturing, kaizen events, and so on.

> It evidently becomes for each man's interest, then, to see that no job is done faster than it has been in the past. The younger and less experienced men are taught this by their elders, and all possible persuasion and social pressure is brought to bear upon the greedy and selfish men to keep them from making new records which result in temporarily increasing their wages, while all those who come after them are made to work harder for the same old pay.
>
> The writer was much interested recently in hearing one small but experienced golf caddy boy of twelve explaining to a green caddy, who had shown special energy and interest, the necessity of going slow and lagging behind his man when he came up to the ball, showing him that since they were paid by the hour, the faster they went the less money they got, and finally telling him that if he went too fast the other boys would give him a licking.

The 300 high-priced workers know, on the other hand, that the cost of poor quality and other wastes comes out of their paychecks as well as their employer's profits and will therefore adopt a zero-tolerance attitude toward any part of their jobs that wastes time, materials, or energy.

Cheap Labor Is Costly

Basset (1919, 64) explained clearly why high-priced workers prove less expensive in the long run than cheap ones.

> We all know that cheap labor is not cheap; paid cotton-pickers have proved cheaper than slaves—although it took a long time so to convince the South, because they never reckoned the expense of idle slaves. In any operation in which the material costs are high as compared with the labor costs, the highest possible pay is the cheapest if it results in savings of material, or in a fine product, or in both.

Basset's observation that paid workers were more cost-efficient than slaves for cotton harvesting supports this chapter's later material on the role of automation in the eradication of slavery. Even in the absence of automation, paid workers have an incentive to discover ways to harvest cotton more quickly while slaves and other unpaid laborers (such as those performing robot or corvée) do not.

Ford (1930, 53), who had by then created the most productive manufacturing establishment on earth, added,

> Good workmanship has to be paid for, and good workmanship is cheap at almost any price. It is simply a waste of time and money to erect an elaborate manufacturing equipment and then expect that it can be run by low-paid men.

Ford (1926, 119–121) paid premium wages to the crews of his ships that carried iron ore on the Great Lakes because cheap crews would not care how long their ship remained in port, and the cost of the delay could far exceed a year's pay. The ships were rarely in port for more than a day when Ford's high-priced crews ran them; their goal was to get the cargo aboard and then get it moving as rapidly as possible. High-priced workers and their employers understand that the cost of the waste comes from their wages and profits, respectively, and will therefore remove the waste as rapidly as possible.

What about Piece Work?

Sweatshop owners of the nineteenth and early twentieth century knew that low pay did not earn the loyalty or commitment of their workers, so some used piecework as an incentive to work harder. The workers were paid according to the number of items, such as garments, they produced. Perhaps the workers in this nineteenth-century sweatshop (Figure 4.1) are indeed working as hard as they can, but the poor organization—the 5S workplace organization approach is clearly absent here—means they are far less productive than they could be. In addition, if they are paid by the piece, quality is unlikely to be among their priorities.

Barboza (2008) reports, however, that sweatshop labor is still common in the People's Republic of China (PRC). One of the sweatshops in question, in fact, paid piece rates but never told the workers how much. High-priced American workers, in combination with intelligent management, ought to be able to do to the offshore sweatshops what automation or even motion-efficient paid manual labor could have conceivably done to slavery in the antebellum South and without the need to fight the Civil War.

Can We Compete with PRC Automation?

The PRC is of course also building factories with automation and may even be introducing lean manufacturing while increasing the pay of some of its workers. This creates the worst-case scenario in which this dangerous geopolitical rival, with its far larger population, implements the productivity improvement methods this book will cover in detail later. This does not, however, get rid of the Pacific Ocean and the associated transportation costs and inventory carrying costs of exporting PRC-made goods to the United States. This means that while the PRC would win a contest between its own high-priced workers and automation and those of the United States at home for the same reason—the transportation and inventory carrying costs of shipping our goods to the PRC—it would lose that contest in North America. It is for the same reason that, despite Japan's own high-priced workers, extensive automation, and world-class quality and productivity

Figure 4.1 Nineteenth-Century Sweatshop. Riis, Jacob A. circa 1889. Workers in a sweatshop in Ludlow Street Tenement, New York City. Note the garments and/or materials on the floor, and the man in the center working in an awkward position (United States Library of Congress's Prints and Photographs Division, digital ID cph.3a24271. Public domain due to age.

methods, many Japanese cars that are sold in North America are built in North America. The Japanese do not want to incur the transportation and carrying costs, along with the risk of exposure of vehicles to spray from seawater on their way to the United States.

We have seen so far, and this chapter will elaborate further later, that high-priced people usually prove more cost-effective than cheap ones. The problem is that the executives and managers who offshored American jobs were apparently incapable of understanding this even though Ford's publications were as accessible to them as they are to us. They were also incapable of looking past the dysfunctional financial metrics this chapter will discuss later. The basic issue relates to what Emerson (1924, 92 ff.) called near common sense and supernal common sense as discussed previously; the next section will go into extensive detail.

Near Common Sense versus Supernal Common Sense

Emerson described previously how Daniel Webster apparently opposed the expenditure of money to extend mail service to the Pacific Coast even though the long-term benefits would have far outweighed the $50,000 involved. He adds additional examples of near common sense, i.e., focus on short-term costs and benefits, and supernal common sense, which accounts for long-term results as well as risks and opportunities that are not immediately obvious. Near common sense is why Nevada was once essentially a large-scale boom town that returned mostly to the desert when the gold and silver ran out.[2]

> Does the American paranoiac differ much from the American State of Nevada which a generation ago, in its golden youth, took $300,000,000 in gold and silver from the ground, exported it all for transitory equivalents, and then lapsed into a sparsely settled desert waste? (Emerson, 1924, 97)

Recall that the people who really make money from a gold rush are not the prospectors, but rather the general store proprietors who sell equipment, and not always of the best quality, to the prospectors. The general store (Figure 4.2) may have to close when the gold runs out, but it can move elsewhere, and the factories that supply it will be around long after the boom town has returned to the desert. Boggan (2015) reports that during the California Gold Rush, a large breakfast could cost $43 *in the money of 1850*, and a buttered slice of bread cost two (silver) dollars. "While some miners did strike it rich in the early days, those that made most money were the ones who 'mined

Figure 4.2 A General Store. Walker Evans, for the Farm Security Administration. 1936. "General store interior. Moundville, Alabama, USA,". A general store of the California Gold Rush would have probably looked similar minus the Coca-Cola poster (as the drink was introduced in 1886); note the factory-made lanterns, dishes, soap, and other products without which the prospectors could not have stayed in business (public domain as the work of a US Government employee)

the miners.'" One woman who did no prospecting at all made $18,000, again in the money of the mid-nineteenth century, by selling pies to those who hoped to strike it rich. More to the point, pickaxes that were probably made in factories back east cost $1,500 in today's money. Philip Armour (of Armour Foods) got his start by selling meat to prospectors, Levi Strauss recognized the prospectors' need for sturdy clothing, and the founders of Wells Fargo established banking services in San Francisco.

The prospectors' "near common sense" told them they could strike it rich by panning or digging for gold, and some did, but the "supernal common sense" of Philip Armour, Levi Strauss, and Henry Wells and William Fargo is why their businesses are still around 170 years later. It would come as no surprise if the manufacturers and sellers of the massive computer servers that people use to "mine" Bitcoin and other cryptocurrencies, and the power companies that sell the enormous amount of electricity necessary to do this, end up similarly as the only participants to make money in the long run. Cryptocurrency is otherwise a zero-sum game in which some speculators get rich while most others suffer enormous losses. Bitcoin is in fact down about 70 percent (as of October 2022) from its high in November 2021, and similar speculation in Dutch tulip bulbs during the seventeenth century comes to mind.

Emerson (1924, 97–98) then goes on to compare the near common sense of selling raw materials and commodities, and even precious metals, to the supernal common sense of selling the same materials only after the addition of value to them.

> Switzerland was to Europe what the western deserts were to North America, a region destitute of national resources, but for centuries the canny Swiss marketed the fighting skill of their sons, who hired out in companies as guards for kings like Louis the XVI of France, or as gateway guards for private palaces, until in French the word "suisse" has become to mean "front-door custodian."
>
> The Swiss also began to market little blocks of lumber for their weight in silver (after they had carved them by hand and brain skill). They imported raw materials from $20 a ton up, and they exported them again as watches worth from $32,000 to $16,000,000 a ton, the difference between import value and export value being Swiss brains and handicraft. A very high order of supernal common sense animates the Swiss.

This is also why Japan and Israel, both of which are very poor in natural resources, have some of the highest per-capita incomes in the world as does Switzerland. The same theme appears in science fiction stories such as Gordon R. Dickson's Dorsai and Isaac Asimov's Foundation series. The Dorsai, who are thinly disguised Swiss mercenaries with spaceships—they even live in cantons—inhabit a resource-poor world and hire themselves out to other planets. Asimov's Foundation was created on a resource-poor planet, so its people had to rely on technology and the creation of value to gain a dominant position. Supernal common sense therefore supports the transformation of raw materials into high-value manufactured goods. Near common sense, as supported by dysfunctional financial metrics, supports offshoring and/or cheap labor.

Financial Metrics: The Road to Ruin

It is a well-known principle in corrective and preventive action (CAPA) that, if we ask the wrong question, we will get the wrong answer. If we rate people's performance on the wrong metrics, and financial metrics are often among them, we will similarly get very dysfunctional results that can include outsourcing of manufacturing capability at an actual loss. Many cost accounting systems end up as suicide pacts for the organizations that rely on them for executive decision-making, including make versus buy or make versus outsource.

Traditional cost accounting methods that conform to generally accepted accounting practices are, while mandatory for financial and tax reporting, utterly ruinous when it comes to managerial decision-making. Peters (1987, 588) quotes H. Thomas Johnson's statement, "Cost accounting is the number one enemy of productivity" while Panchak (2014) asks, "Did Finance Gut Manufacturing?" The answer is yes because financial metrics often lead to dysfunctional and counterproductive actions. Sekora (2014) elaborates, "The disease killing America's economic health is financial-based planning, and one of the symptoms of this ongoing disease is the loss of the U.S. manufacturing base."

Henry Ford (Ford and Crowther, 1922), whose industries once made the United States the wealthiest and most powerful nation on earth, stated,

> And that is the danger of having bankers in business. They think solely in terms of money. They think of a factory as making money, not goods. They want to watch the money, not the efficiency of production … The banker is, as I have noted, by training and because of his position, totally unsuited to the conduct of industry.

This ties in with Emerson's discussion of near common sense, i.e., a short-term focus on immediate results, as opposed to supernal common sense which is a long-term outlook. This section will show, however, that there are situations in which financial metrics that treat overhead and even labor as marginal or differential costs are not even near common sense because they are wrong from even a short-term perspective.

Be Careful What You Wish; You Might Get It

Jacobs's (1902) *The Monkey's Paw* is a terrifying horror story in which a talisman gives its owner three wishes and grants them literally. "'The first man had his three wishes. Yes,' was the reply, 'I don't know what the first two were, but the third was for death. That's how I got the paw.'" This is only a story, but its real-world counterpart undermined dangerously the readiness of the Royal Navy during the second half of the nineteenth century. Nobody had dared to challenge seriously the Royal Navy since the Napoleonic Wars, and the result was complacency that reached the point where captains were assessed more on the smart appearances of their ships than on their gunnery. Smoke from gunnery practice marred the ships' immaculate appearances so some captains threw practice ammunition overboard rather than fire it. The *Times Sunday Special* (1898) reported,

> Compared to Our Gunners, Those of England are Woefully Lacking in This Branch of Naval Warfare—Gunnery Practice in the Royal Navy … the subjects of the Queen are learning to their dismay that little, if any, importance is attached to the training of the gunners in what is supposed to be the first navy in the world.

This article adds,

> At present in the Mediterranean far more trouble and time is expended in filing the chase of a gun bright and burnishing it (which is absolutely contrary to regulations) than to insuring that the men are well trained in the use of it.

It is probably fortunate for the Royal Navy that the First World War started in 1914 rather than 1898 when a numerically inferior but far more competent enemy might have exploited these deficiencies decisively. As matters stood, the German capital ships at Jutland (1916) scored a significantly higher proportion of hits per shell fired than their British counterparts. Dogger Bank (1915) was even worse. If we don't count hits scored by the British on a crippled heavy cruiser (*Blucher*) after it was unable to defend itself, the German capital ships scored three major-caliber hits in exchange for each received from their British counterparts. This suggests that the priorities, as reflected by performance measurements, in the Imperial German Navy differed significantly from those in the Royal Navy.

Modern management expresses similarly its wishes through financial performance measurements, and the results can be equally horrifying to the business's stakeholders. The consequences can even include bankruptcy and the relegation of a once-great business to the history books.

How Dysfunctional Metrics Brought Down W.T. Grant

W.T. Grant was once a major retailer throughout the United States. Its bankruptcy in 1976 was the second biggest in the country's history at the time and resulted apparently from dysfunctional performance metrics. Sales representatives were rated on their sales volume and credit sales counted as such. They were also authorized to issue credit cards to customers, and they did so without regard to the customers' ability to pay. Glasberg (1989) quotes a no-longer-available *Business Week* article in which a W.T. Grant manager complained, "We gave credit to every deadbeat who breathed."

Unsalable Inventory and the Laxian Key

Robert Sheckley's (1968) science fiction story *The Laxian Key* features an alien device whose human discoverers turn on, whereupon it sucks energy out of nearby power lines to produce a gray material. The power company demands that the machine's owners pay for the enormous amount of power it consumes, whereupon they discover they cannot turn it off without something called a Laxian Key. When they discover that the gray material, which is known as Tangreese, is edible on an alien planet, they try to dispose of the machine there. The angry inhabitants, who are already buried in so much Tangreese that they cannot even begin to consume it, demand that the humans get the machine off their planet. They add, however, that if the humans can find a Laxian Key, they can name their price for it. Overproduction is, of course, one of the Toyota Production System's (TPS) Seven Wastes.

Dysfunctional performance metrics can similarly encourage a factory to produce so much inventory that it cannot even begin to use it all. Suppose for example that labor costs $25 an hour, or $200 a day. It is therefore "obvious" that, if each worker can produce 200 rather than 100 units a day, the labor cost per unit will drop from $2 to $1 and may even yield a favorable accounting variance, at least on paper. If however the downstream process requires 100 units a day, the company will simply tie up money (in the form of the material costs of making each unit) in inventory it will never use. Absorption of overhead costs is yet another motive to produce items regardless of whether they will ever be needed, and so are incentives to keep equipment busy to distribute the capital costs over as many items as possible.

If we remember that overproduction is one of the Toyota Production System's Seven Wastes, and it ties up money in unusable and possibly unsalable inventory (another of the Seven Wastes), this is a perfect example. It might be useful to have production workers and managers read

Sheckley's story (this requires perhaps 10–15 minutes) and then refer to inventory as "Tangreese" rather than taking it for granted as many organizations do.

Franklin (1758) wrote of this,

> [Poor Richard] means, that perhaps the cheapest is apparent only, and not real; or the bargain, by straightening thee in thy business, may do thee more harm than good. For in another place he says "Many have been ruined by buying good penny worths."

"Straighten" does not mean to make straight; it is an eighteenth-century English variant of straiten, i.e., to constrain or confine in the sense of a strait jacket, or a narrow waterway called a strait. One can do a lot more with ready cash, which is a quick asset in accounting, than with inventory which, while nominally a short-term asset, cannot always be liquidated. Ford (Ford and Crowther, 1922) liquidated his inventory to make cash more readily available.

> We had been carrying an inventory of around $60,000,000 to insure uninterrupted production. Cutting down the time one third released $20,000,000, or $1,200,000 a year in interest. Counting the finished inventory, we saved approximately $8,000,000 more—that is, we were able to release $28,000,000 in capital and save the interest on that sum.

Marginal Revenues, Costs, Profits, and Sunk Costs

In this case, the "Laxian Key" necessary to turn off the overproduction consists simply of recognition of the fact that accounting costs do not reflect everyday reality. Per Carr (1987, April 28), the prime cost of goods sold consists of direct material and labor and the addition of (1) factory overhead, (2) costs of selling, and (3) general and administrative costs yields the full cost, total cost, or "cost to make and sell."

Anthony and Reece (1983, 551–553) explain, "The full cost of a cost objective is the sum of its direct costs plus a fair share of applicable indirect costs." Illustration 17-1 of this reference explains,

1. Prime cost = direct material and labor cost
2. Full production cost = prime cost + overhead cost
3. Full cost = full production cost + selling cost + general and administrative cost

The truth is however that the overhead, selling, and general and administrative costs are *sunk costs*. This means we incur them regardless of how many units we make and sell, and the decision to make or not make does not affect these costs in the least. Anthony and Reece (1983, 712) make it emphatically clear that "a sunk cost is not a differential cost" and include an example related to a machine's depreciation that underscores the irrelevance of sunk costs to current decision-making. Even labor, which everybody "knows" to be a direct variable cost, also is essentially a sunk cost unless we are paying overtime. This is because we pay the workers to be present eight hours a day regardless of whether we use them. This leaves the cost of materials, and energy if applicable, as the only genuine incremental costs of creating another unit of production.

Suppose again that labor costs $200 per worker per day regardless of whether we keep the person busy for the full day and also that the marginal cost of another unit's production is $3 in materials. An internal or external customer needs 100 units a day @$5.00. This is where marginal revenues, costs, and profits come into play.

- The marginal revenue is the amount of money realized from the sale of an additional unit, which is $5.00.
- The marginal cost is the cost of making the additional unit, which is $3 for materials.
- The marginal profit is the marginal revenue minus the marginal cost, and it is the actual cash flow that takes place. This comes to $2.00.
- The actual cash flow is therefore the volume times $2.00 minus the $200 fixed cost for the labor.

Traditional cost accounting would conclude from this, if we make 100 parts a day to meet the downstream demand, their unit cost is $2 in labor plus $3 in material or $5 which equals the sale price. This matches the cash flow of 100 @$2.00 marginal profit minus the $200 fixed labor cost, so we break even.

If we make 200 a day, the accounting system computes a cost of $1 in labor plus $3 in material, or $4 which is less than the sale price. The practical truth is however something very different. Remember that we pay $200 a day for the labor regardless of how many items we make, so the actual cash flows are as follows:

- 100 @$5.00 minus $200 labor and $300 material = 0 (breaks even as shown previously).
- 200 a day: 100 @$5.00 minus $200 labor and $600 material = ($300) either as an outright loss if the inventory—think of Sheckley's "Tangreese"—can never be used, or $300 tied up as inventory. We have alternatively the original 100 @$2.00 marginal profit less $200 in labor and also less $300 for the material we tie up in inventory, for a cash flow loss of ($300). The sole effect of this decision is therefore to turn cash, a quick asset, into inventory. The fact that inventory is, at least on paper, a short-term asset rather than one of the TPS's Seven Wastes disguises its real nature the same way menthol disguises the harmful and irritating nature of cigarette smoke; it makes it easier to inhale.
- Carr (1987, May 5) adds overhead costs in the context that if production exceeds sales, costs are not really absorbed; they are instead stored in the inventory until the goods are sold. The fact that inventory is, at least on paper, a short-term asset gives these costs a perfect place to hide. In other words, the "near common sense" of overproducing to absorb costs does not even accomplish that goal.

The correct course of action is therefore ideally to increase the downstream demand for the item to 200 a day or, if this is not possible, assign half of the worker's time to the production of something for which there is a relatively immediate need. If that is not an option either, then the worker can be assigned to continual improvement activities, but the organization should not make an unsalable inventory.

The issue of marginal revenues, costs, and profits also applies to decisions as to whether to accept an order at a price that will result in a loss, at least on paper. Ford (Ford and Crowther, 1922) wrote of this, "He can take the direct loss on his books and go ahead and do business or he can stop doing business and take the loss of idleness." The loss of idleness is the marginal profit the organization foregoes because the transaction appears to lose money according to standard costs for labor and overhead. Suppose that, under the conditions described above (labor @$200 per worker per day and materials @$3 a unit), an external customer offers to buy 100 units a day for

$3.50 each. Remember that we can make 200 a day without paying for overtime, but our internal customer requires only 100.

The cost accounting system will doubtlessly tell us that, as the unit cost consists of $1 in labor and $3 in materials, we will lose 50 cents on each item and should therefore decline the offer. If however we look at the actual cash flow, remember that we pay the worker $200 a day regardless of whether we use his or her labor, and it does not cost more in labor to make 200 units instead of 100. Only if we must pay overtime does labor become a genuine variable cost. The only true variable cost is therefore the $3 for material which means we get 50 cents per unit, or $50 a day, we would not have otherwise had. Note by the way that the cost accounting system will itself eventually recognize this marginal profit when it subtracts actual expenses from the revenues, but it does not recognize them at the standard cost level that applies a certain labor and overhead cost to each item.

Goldratt (1992, 311–313) describes a similar situation in which a customer offers $701 per unit for a product, and the plant manager tells the sales manager to take the deal even though the latter says the customer is asking for the product to be sold for practically nothing. This might again be true on paper if the accounting system factors in labor, overhead, and other fixed costs. The plant manager, however, has unused capacity which means the marginal cost of filling the order is only the $344 material cost. Traditional cost accounting would have therefore foregone a chance to use the plant's excess capacity to bring in real money.

Transfer Pricing Traps

An item's transfer price is the price charged to an internal customer by an internal supplier in the same company. Carr (1987, May 12) warns that problems with negotiated internal transfer prices can arise when the end seller wants to exploit a market opportunity but needs the cooperation of another division in the same company. The latter may not want to adjust its output if doing so will result in unfavorable accounting variances, even if the overall transaction is in the best interests of the company. This reflects the lesson taught by the Goldratt example in which a purported loss, at least according to valuations by the accounting system, is actually a positive cash flow and eventual profit.

A simple example might be a transfer price that is based on cost, or cost-plus, with "cost" including sunk expenditures such as overhead, and even labor if overtime is not being paid. The transaction might even deliver a marginal profit to the internal supplier as well as its internal customer, but the former might refuse to take it because it loses money, at least ostensibly, on paper. This is known as *suboptimization*, a form of "near common sense" in which a local optimum decision is not in the long-term best interest of the entire organization. Suboptimization is the easily foreseeable result of dysfunctional performance metrics.

How to Outsource Manufacturing at a Loss

The issue of managerial economics carries over into decisions as to whether to make or buy an item. Suppose for example labor costs $1,000 a day, and the accounting system assigns a 400 percent overhead burden to make the purported labor cost $5,000 a day. Management decides accordingly to pay an offshore manufacturer $2,000 a day, which saves $3,000 on paper but squanders in practice $1,000 (the $2,000 we pay the supplier minus the $1,000 we don't pay our own workers) because the $4,000 overhead cost is actually a sunk cost. We pay it regardless of whether we use the associated labor and tool hours, and some accounting systems also allocate overhead costs to tools. The only way to get rid of the overhead is therefore to liquidate the plant and equipment, and probably at a loss because if we can't get a profit from it, then neither can any prospective buyer. A competitor who does not rely on dysfunctional financial metrics might, on the other hand, "take it off our hands" and then use it to compete very effectively in the same market.

Overhead includes, among other things, rent paid for production space, utilities, insurance, depreciation, property taxes, and health insurance (Reynolds, Sanders, and Hillman, 1984, 821); the only variable cost appears to be overtime pay. Peters (1987, 587–588) explains why overhead is totally irrelevant to actual costs: "You can't shut the heat off around one idle machine," and adds an outsourcing example in which the business not only fails to get rid of its own overhead costs but adds new costs associated with outsourcing.

If we go back to the Goldratt example, in fact, an intelligent domestic competitor with excess capacity might offer similarly to sell the product to us for less than what it costs (on paper) for the competitor to make. That is, the competitor might rightfully ignore its own overhead costs and other sunk costs to put its unused capacity to work and persuade us to discharge our own workers and sell off the associated plant and equipment in the bargain. The competitor will then have us at its mercy because it now controls the means of production. While nothing in this book constitutes legal advice, I am not even sure that anybody could file a complaint about predatory pricing (selling at a loss to drive competitors out of business) under antitrust laws because the competitor in question could prove it is actually making money and reporting taxable profits it would have otherwise not realized.

Ford (Ford and Crowther, 1922) told us 100 years ago,

> The foremen and superintendents would only be wasting time were they to keep a check on the costs in their departments. There are certain costs—such as the rate of wages, the overhead, the price of materials, and the like, which they could not in any way control, so they do not bother about them. What they can control is the rate of production in their own departments.

While the cost accounting system may allocate overhead to the product, workers, and/or machines, it is relatively meaningless on a practical basis and attempts to apply meaning to it lead frequently to the wrong decisions.

Dysfunctional Purchasing Incentives

Purchasing incentives that encourage the purchase of items for which there is no immediate use is similar in principle to financial metrics that encourage the generation of unnecessary inventory. Suppose, for example, the purchasing department is rated on the price it pays for a given commodity, and it can get a lower price by buying more. Goldratt (1992, 37) described a situation of this nature.

> They're out there renting warehouses to store all the crap they're buying so cost-effectively. What is it we have now? A thirty-two-month supply of copper wire? ... They've got millions and millions tied up in what they've bought—and at terrific prices.

Franklin (1758) warned explicitly against this practice:

- "You call them goods; but, if you do not take care, they will prove evils to some of you. You expect they will be sold cheap, and, perhaps, they may for less than they cost; but, if you have no occasion for them, they must be dear to you."

- "Buy what thou hast no need of, and ere long thou shalt sell thy necessaries."
- "Many have been ruined by buying good penny worths."

Henry Ford, who cited Franklin as an influence on his own business practices, elaborated on Franklin's principles considerably (Ford and Crowther, 1922, emphasis is mine):

> But we learned long ago never to buy ahead for speculative purposes. When prices are going up it is considered good business to buy far ahead, and when prices are up to buy as little as possible. It needs no argument to demonstrate that, if you buy materials at ten cents a pound and the material goes later to twenty cents a pound you will have a distinct advantage over the man who is compelled to buy at twenty cents. But we have found that thus buying ahead does not pay. It is entering into a guessing contest. It is not business. If a man buys a large stock at ten cents, he is in a fine position as long as the other man is paying twenty cents. Then he later gets a chance to buy more of the material at twenty cents, and it seems to be a good buy because everything points to the price going to thirty cents. Having great satisfaction in his previous judgment, on which he made money, he of course makes the new purchase. Then the price drops and he is just where he started. We have carefully figured, over the years, that buying ahead of requirements does not pay—that the gains on one purchase will be offset by the losses on another, and in the end we have gone to a great deal of trouble without any corresponding benefit. Therefore in our buying we simply get the best price we can for the quantity that we require. *We do not buy less if the price be high and we do not buy more if the price be low.* We carefully avoid bargain lots in excess of requirements. It was not easy to reach that decision. But in the end speculation will kill any manufacturer. Give him a couple of good purchases on which he makes money and before long he will be thinking more about making money out of buying and selling than out of his legitimate business, and he will smash. *The only way to keep out of trouble is to buy what one needs—no more and no less.* That course removes one hazard from business.

This reinforces the takeaway that dysfunctional performance metrics can result in the purchase of items for which there is no immediate need and which might even become obsolete because of changing business needs.

Total Cost of Ownership and Toyota's Seven Wastes

This book mentioned previously the total cost of ownership (TCO), which includes hidden or other costs that are not evident in the ostensible purchase price. Consider for example hexavalent chromium (chromium that makes six connections to other atoms) that is used in electroplating, wood preservation, pigmentation, fungicidal compounds, and other applications. A webinar (Mayo, 2022) on the hazards associated with hexavalent chromium including carcinogenetic effects, respiratory effects, eye hazards, and skin irritation was enough to give me a real distaste for chromium in its hexavalent state.

Now suppose a process owner must choose between a hexavalent chromium compound and a less hazardous substitute with equal performance but a much higher price tag, even after the process owner accounts for the cost of chromium waste disposal in accordance with environmental regulations. This makes hexavalent chromium the "obvious" choice, but when one considers all the additional costs related to mandatory air quality monitoring, personal protective equipment

(PPE), safe handling of contaminated PPE, avoidance of cross-contamination of street clothing by contaminated PPE, cleanup, medical surveillance, a respiratory protection program that meets the requirements of Code of Federal Regulations (CFR 1910.134), and other considerations,[3] the total cost of ownership (or total cost of use) of the hexavalent chromium compound may far exceed its ostensible price tag plus the relatively obvious cost of proper waste disposal. This reinforces the difference between the near common sense that looks only at the obvious costs and the supernal common sense that accounts for the big picture. The lesson that we can easily incur costs that don't appear in the obvious places carries over into costs related to offshoring.

Moser (2015) shows how costs related to offshoring constitute one or more of the Toyota Production System's Seven Wastes. These include inventory, which is unavoidable when one brings in container ships rather than just-in-time quantities, the cost of transportation itself which often includes empty or partially empty ships returning to their points of origin without the American-made goods they should carry, costs of packing and unpacking, and high defect levels that can easily hide in large shipments.

While Henry Ford was able to operate a supply chain that Norwood (1931, 20–24) depicted as a continent-spanning conveyor, with railroad shipments that ran like clockwork and with contingency plans to redirect shipments in response to force majeure, delivery from offshore sources is far less reliable. Ford himself wrote of this issue (Ford and Crowther, 1922),

> If transportation were perfect and an even flow of materials could be assured, it would not be necessary to carry any stock whatsoever. The carloads of raw materials would arrive on schedule and in the planned order and amounts, and go from the railway cars into production. That would save a great deal of money, for it would give a very rapid turnover and thus decrease the amount of money tied up in materials. With bad transportation one has to carry larger stocks.

Ford means it is necessary to carry safety stock (aka inventory) for protection against supply chain interruption if delivery is not reliable. Delivery becomes more reliable over shorter distances and with fewer changes in transportation modes (e.g., container ship to dock, dock to railcar or truck, and so on).

Waste in Trucking Hurts Drivers and Makes Just-in-Time Impossible

The ideal just-in-time situation is simply not achievable when one has to deal with container ships with uncertain departure times (e.g., due to logistics issues in the country of origin), uncertain sailing times, and uncertain arrival times plus issues associated with dock capacities in the United States, availability of trucks or rail cars, and so on. Truck drivers have in fact complained about being told to wait for up to eight hours, and without pay, at ports in Southern California (Kay, 2021). These delays have in turn caused truck drivers to miss drop-off times at warehouses (and presumably other destinations like factories that need their cargoes) which is of course inconsistent with just-in-time production control systems.

Sainato (2021) adds that one driver reports, "He has waited up to 36 hours, with typical wait times of several hours." Waiting is of course another of the TPS's Seven Wastes, and matters become even worse when drivers are paid by the mile. The driver added of this, "If the wheels aren't turning, the driver isn't earning." It would therefore come as no surprise that drivers who are told to wait without being paid will refuse to perform this kind of work, seek instead domestic routes where these delays are not going to be a problem, or even get out of the business entirely. It must also be pointed out that if, for example, 25 percent of the driver's time and the truck's time is wasted in this manner, we need 33 percent more drivers and trucks to do a given amount of work. The cost of this excess must of course be borne by truck drivers in the form of lower wages or unpaid work, customers in the form of higher prices, and businesses in the form of lower profits. Truck drivers have every right to complain about this, and so do supply chain partners that are affected by these inefficiencies.

It is highly doubtful, by the way, that a driver would have had to wait more than once in any supply chain managed by Henry Ford. Ford bought out the Detroit, Toledo & Ironton Railway because he couldn't tolerate unreliable deliveries in what was effectively a just-in-time production control system, and he didn't like the high prices he had to pay or the low wages received by railroad employees. Ford wrote explicitly (Ford and Crowther, 1922),

> The trains must go through and on time. The time of freight movements has been cut down about two thirds. A car on a siding is not just a car on a siding. It is a great big question mark. Someone has to know why it is there. It used to take 8 or 9 days to get freight through to Philadelphia or New York; now it takes three and a half days.

A truck sitting idle, with its driver going unpaid and its engine possibly idling and wasting fuel, also ought to be a great big question mark. Ford would have doubtlessly asked why the driver had to wait and why the truck (a capital asset itself) was not doing productive work and then ensured that it never happened again.

All of this reinforces the conclusion that the total cost of ownership, or total cost of use, of imported goods can be far higher than the ostensible price tag that makes these so attractive to purchasing departments.

Book Value Is Not Real Value

The book value of an asset is important in cost accounting because it reflects (1) the depreciable amount that remains in the asset and (2) the value of the asset if it is sold. Suppose for example that a machine was purchased for $500,000, and $200,000 has been written off as depreciation. $300,000 remains to be depreciated over the accounting life (five years is often used in textbooks), and if the machine is sold for $250,000, there is an immediate $50,000 capital loss.

The machine's value also counts toward the denominator for the calculation of return on assets (ROA) and return on investment (ROI). If the machine depreciates to zero value, there is a strong incentive to not replace it with a superior model because the ROI and ROA will decrease. This means the company may fall behind a more advanced competitor. Ford (1930, 25) said of this issue, "As for the buildings and machinery, they must be valued in dollars according to the meaningless methods of accounting that are required by law. Actually they are worth only what we can do with them." That last phrase, "*They are worth only what we can do with them*" is worth remembering.

The same principle applied to the housing crash of 2008, which resulted because people bought, and often with subprime mortgages, houses they thought would continue to increase in value forever. Dutch speculators thought the same thing about tulip bulbs in the seventeenth century, and the result was equally catastrophic. A house, like a piece of machinery, is worth only what the owner can do with it. This is to live in it without paying rent to a landlord, and avoided rent expense is essentially after-tax income. A house is not an investment to be "flipped" for an easy profit unless one is a builder who can add enough value to make it more attractive to a buyer.

Nor is it a source of money as might be obtained through a home equity line of credit (HELOC). Financial talk show host Dave Ramsey (2021) has plenty to say about HELOCs, and none of it is good.

Costs of Foregone Opportunities

Opportunity costs, or the money we don't make by foregoing opportunities, are *totally invisible to the cost accounting system* because we cannot write them off on an income tax return or report them on an income statement. Goldratt's Theory of Constraints warns, for example, that time lost at the capacity-constraining production resource is lost forever. Suppose it costs $1 to rework a nonconforming part in the constraint, and the end product sells for a $5 profit. The accounting system recognizes the $1 cost of the rework, but the total cost is $6; the cost to rework the part plus the foregone marginal profit of making another item. This book will treat the issue of opportunity costs in detail later.

This section has shown that cost accounting metrics have their place, and conformance to generally accepted accounting practices is mandatory for the preparation of financial reports and tax returns. Their deployment to the shop floor can, on the other hand, be utterly ruinous to actual business performance. The next section will go further into the false economy of cheap labor.

Slavery, Robot, and Corvée as Free Labor

If low-wage labor is good, then free labor must be better. Slavery is self-explanatory, while robot (the origin of the word for a mechanical worker) and corvée were taxes payable in labor rather than money. Medieval or even relatively modern overlords, noting that serfdom and robot prevailed in Austria, France, and Prussia through the eighteenth century and Russia into the nineteenth, might demand that their subjects work for them for a certain number of days per month. Cotton is apparently still harvested by hand in Uzbekistan in the twenty-first century with unpaid labor. Slavery, robot, and corvée have however been abolished throughout the advanced and civilized world, and not just for human rights reasons. Widespread opposition to slavery began when industrialization made it uneconomical, just as opposition to the use of animal fur for clothing began when artificial substitutes became available.

Emerson (1924, viii) described exactly why automation makes slavery, robot, corvée, and even animal power uneconomical. He determined the cost of horsepower by source:

1. Human: $54,000 per year of 7,500 hours, and this was in the money of the early twentieth century.
2. Small gasoline engine, $300 per year.
3. Large power engine, $20 to $200 per year.

Emerson wrote of this, "Man power costs therefore from 135 to 1350 times as much as uncarnate [not made of flesh, i.e. mechanical] power" and "Thirty men, as men work, will yield 1 horse power of energy each hour, but so will 1 to 5 pounds of coal." This means that almost any powered machine is far cheaper to operate than it is to feed, even at a subsistence level, the number of slaves, serfs, or other unfree workers necessary to perform the same job. A machine is similarly far cheaper to operate than horses, donkeys, mules, and other animals that can be trained to work solely for food, shelter, and veterinary care. The Ford Motor Company's tractors revolutionized agriculture roughly a century ago, and nowhere in the United States, other than on Amish-owned farms, is farm equipment still pulled by animals.

The only technical barrier is to convert the mechanical energy to get it to pick cotton, move water from a lower to a higher elevation, dig a ditch, or whatever else is required, and engineers have overcome this barrier routinely. The harvest of cotton once required some skill to avoid damage to the bolls, but a modern cotton harvesting machine automates this skill to do more work than a thousand hand laborers. The cost of these laborers' basic subsistence under slavery, robot, or corvée would far exceed the cost of the machine's fuel, maintenance, and capital costs along with the high wages that can be paid to its operator and the mechanics who keep it in good working order.

Aristotle Predicted That Automation Would Abolish Slavery

Aristotle (350 BCE) predicted accurately that automation would make slavery obsolete, and automation was known to the ancient Greeks. Their myths and legends included mechanical men, whom we would now call robots, who assisted the god Hephaestus in his workshop. Hephaestus also built a giant bronze assault robot called Talos whose function was to protect the island of Crete from invasion, and he appears in the movie Jason and the Argonauts (1963) courtesy of stop-motion animator Ray Harryhausen. Daedalus built what we might now call hang gliders to escape from King Minos.

The Greeks did not content themselves, however, with telling what we might call science fiction stories today. They acted on these stories by inventing some very ingenious devices:

- The Diolkos, or portage machine, allowed boats to be moved across the Isthmus of Corinth so they would not have to circumnavigate the Peloponnese peninsula. It was essentially a railway, which evolved independently centuries later from the tramways that miners used to facilitate the movement of ore cars.
- The canal lock, as installed in an ancient Suez Canal by Ptolemy II (the son of one of Alexander the Great's successors).
- Archimedes' screw can transport water from a lower elevation to a higher one. It can be powered by wind, which offers a labor-free way to obtain large quantities of water for irrigation and other purposes. Turbines and screw propellers evolved from this concept.
- The water wheel provides enormous amounts of mechanical energy for milling grain, crushing ore, and similar jobs.
- The chain drive that now appears in motorcycles and bicycles appeared originally in the Greek polybolos (poly = many, bolos = missile), a self-loading ballista.
- The aeolipile, as invented by Heron of Alexandria, proved that steam could do mechanical work. Heron used this to operate automatic doors in a temple.

Many of these inventions postdated Aristotle (Figure 4.3), but the Greeks obviously knew even during Aristotle's lifetime that natural forces such as water, air, and solar power could do far more labor than humans or domestic animals.

Aristotle (350 BCE) not only predicted that the machines in question would make slavery obsolete but also underscored the clear difference between instruments of production and finished goods (emphasis is mine).

> Thus, too, a possession is an instrument for maintaining life. And so, in the arrangement of the family, a slave is a living possession, and property a number of such

Figure 4.3 Plato and Aristotle. Raphael, "School of Athens," Plato is on the left, and Aristotle is on the right. The bound books are anachronisms, as bookbinding would not be invented for a few centuries; Plato and Aristotle would have read and written scrolls (public domain due to age.)

instruments; and the servant is himself an instrument which takes precedence of all other instruments. For if every instrument could accomplish its own work, obeying or anticipating the will of others, like the statues of Daedalus, or the tripods of Hephaestus, which, says the poet, "of their own accord entered the assembly of the Gods"; *if, in like manner, the shuttle would weave and the plectrum touch the lyre without a hand to guide them, chief workmen would not want servants, nor masters slaves.* Here, however, another distinction must be drawn; the instruments commonly so called are instruments of production, whilst a possession is an instrument of action. *The shuttle, for example, is not only of use; but something else is made by it, whereas of a garment or of a bed there is only the use.*

The possessor of the instruments of production, i.e., factories and their equipment, is clearly in a much better economic position than the possessor only of instruments of use, i.e., finished goods from factories. Most instruments of use, including nominally durable goods, eventually wear out and require replacement, but the instruments of production can always make more.

The Suez Canal: Late and Overpriced with Free Labor

Emerson (1924, 375) describes the construction of the Suez Canal in the mid-nineteenth century, the labor for which was done under corvée. The "free" labor gave the management no incentive to improve the efficiency of the excavation, and jokes about using toothbrushes to scrub floors or teaspoons to dig holes cease to be funny when we realize that this massive job was designed just as badly.

> Forty years ago I watched the workers on the Suez Canal. Many of them were girls, digging up the sand with their bare fingers, scooping it into the hollows of their hands, throwing it into the rush basket each had woven for herself, lifting the baskets to their heads, and carrying the load of 20 to 30 pounds a hundred feet up the bank and dumping it. Panama excavation is being done by steam shovels.

What management got for its "free" labor was a project that finished five years late and with a 166 percent cost overrun. Emerson continues later, "The canal was begun in 1859, estimated to cost $30,000,000 and to be finished in 1864. Its actual cost was $80,000,000 and it was opened in 1869."

There are however pictures of steam-powered machinery working on the Suez Canal, and Mitchell (1895–1896) explains that Ferdinand de Lesseps played a major role in the replacement of laborers with machines.

> M. de Lesseps, always single in his purpose, made the withdrawal of the corvée another stepping stone to success. He procured the reference of the question of damages to the Emperor of the French, who made a generous decision in his favor, far too generous, perhaps, but it enabled the company to introduce machinery in place of hand labor, till the Suez plant excelled that in use in any other part of the world.

Automation Eradicates Slavery and Cheap Labor

Aristotle's principle that automation, and even the kind of automation that the ancient Greeks could only imagine, is the mortal enemy of slavery, and by implication, cheap labor recurred consistently throughout the following centuries. Taylor (1911) wrote, "the one element more than any other which differentiates civilized from uncivilized countries—prosperous from poverty-stricken peoples—is that the average man in the one is five or six times as productive as the other." Ford (1926, 169) added, "the only slave left on earth is man minus the machine."

There was a time when ships, at least in the calm waters of the Mediterranean, were propelled by hundreds of rowers who might be free (as was in the case of the Greek city-states such as Athens where rich and poor rowed side by side) or slaves (as used by the Romans and later the Ottoman Empire). Even if the rowers were not paid, however, they still needed subsistence in the form of food and water, and the quantity that a ship could carry was limited. This meant that Greek triremes and their Persian counterparts could, for example, spend little more than a day at sea. Each sailor on a sailing ship, in contrast, commanded far more motive power from the wind itself than a rower could deploy from his body. A far smaller crew could operate a sailing ship, which eventually made these vessels capable of crossing the Atlantic Ocean and circumnavigating Africa. The galley's sole advantage was that a sailed warship that was becalmed was at the galley's mercy. The introduction of steam power in the nineteenth century eliminated even this last remaining virtue of the galley.

The takeaway so far is that even slaves, the ultimate form of cheap labor, require subsistence which makes it far more cost-effective to pay good wages to a few efficient workers than many cheap or even unpaid inefficient ones. Recall that Emerson (1924, 12) described Helmuth von Moltke's task as follows: "The struggle, before it began, even in its first planning, was to be one of efficiency against inefficiency." A relatively small number of what Frederick Winslow Taylor called "high-priced men" can indeed prove superior to almost any amount of cheap labor.

The same principle carries over into industry where even a limitless supply of unpaid or low-wage labor cannot compete against high-wage labor that has the aid of automation. The women who dug the Suez Canal with their hands still needed sustenance, and it was therefore more cost-effective to pay one high-wage worker to run a steam shovel than to use the "free" labor of 100 or more manual laborers.

False Economy of Cheap Equipment and Training

The supernal common sense of selecting high-priced men, or at least well-trained ones, became obvious in the mid-nineteenth century when British musketeers proved relatively helpless against Afghan skirmishers. Ainsworth (1852) reports that the application of the lessons learned to what the context (1852 and a reference to South Africa) suggests was the Eighth Xhosa War would have saved enormous sums of money as well as, more importantly, British lives (emphasis is mine).

> As a question of economy—and John Bull is very partial to such considerations, even when his honour and glory, the safety of his colonies, or the defence of his own hearth and home are concerned—*if 500 men, armed with the new arm, can do the work of 2000 with the regulation musket, the answer is obvious.* The cost of the war, amounting under the existing system, according to the Times, to 3800 [pounds] per day, but in reality to more than 4000, would, supposing this new system to have been in force, have amounted to only 1000 [pounds] per day.

The "new arm" in question referred to the French Minié Rifle, a muzzle-loader whose hollow-based bullet could be loaded very rapidly but would then expand upon firing to grip the barrel's rifling, and also Prussia's breech-loading von Dreyse needle gun. The same reference cites again the issue of near common sense, i.e., false economy, versus supernal common sense.

> All proposed improvements had, however, to contend in France for a long period of time against the prejudices and parsimony, or rather false economy, of the military and civil authorities, although M. Delvigne, their great advocate, pointed out how the best troops, under the most experienced officers, had been beaten by the rifles of the peasantry of the Tyrol.

This kind of false economy carries over into training—often the first thing to be cut when a business has to reduce expenditures—as well as equipment. Ainsworth (1852) discusses both issues.

The phrase "To save a pound in the expense of the weapon and ammunition, we sacrifice one hundred in the man" is particularly significant because it reflects the near common sense of parsimony in rifle practice that results in potentially fatal disadvantages in battle.

> When it is intimated that the allowance of ball-cartridge to each infantry soldier is only thirty rounds for a whole year's practice, and this to be fired at targets wherein his shots cannot be distinguished, it may be judged whether any emulation can be excited, and what sort of proficiency is likely to be attained. "The lesson," says an old light division officer, "a recruit now learns from his ball-practice, is chiefly what especially hard knocks his musket can give with the wrong end, and too often the object at these parades is only to get them over as quickly as possible." The same authority says: "Probably, on an average, every soldier, by the time he is landed at the Cape, has, from the time of his enlistment, cost at least 100 [pounds]. *Is it not a shame, that from the inferiority of his weapon, and the want of means to make him a decent marksman, he should be rendered comparatively inefficient?* This is a costly mode of proceeding, and the country is deeply interested that the infantry should be no longer crippled by the Ordnance. This lends support to Colonel Chesney's advocacy of an union of the two services. *To save a pound in the expense of the weapon and ammunition, we sacrifice one hundred in the man, besides being disgraced as incompetent.* What, some innocent-minded person would ask, is a soldier trained for if not to use his musket? And which is most valuable, the man or the weapon?"
>
> Some time back it was shown that the knapsack would ride far more lightly on the shoulders if the straps were arranged in a different manner, but the straps and knapsack remain still the same as before. *To save the cost of threepence or fourpence only, two sides of the blade of the British bayonet are made concave, instead of all three, as in the French, and thus two ounces are added to the weight, without any additional strength.*

The same lesson carries over easily into workplaces where the employer is unwilling to make a relatively small capital outlay to increase the productivity of the workers enormously. As but one example, a mechanical floor cleaning machine can be purchased for $10,000 that will do the work of ten people with mops, but there are nonetheless videos and pictures of workers cleaning very large floor areas with mops.

We have seen so far the enormous dangers associated with false economies or, as Emerson put it, near common sense. The next section will address the dangerous paradigm that cheap labor delivers low prices and high profits.

Low Wages Indicate Low Profits and High Prices

Recall that traditional cost accounting treats labor as a direct cost, so less seems better. What really counts, however, is the *labor cost per unit of production*, which is where efficiency principles come into play. Consider bricklaying as practiced around the turn of the twentieth century. This exercise assumes certain costs in the money of 1911 (when home delivery of a loaf of bread cost less than ten cents) but uses actual before-and-after production figures.

- A brick and its mortar cost 0.3 cents.
- The mason is paid 0.3 cents per brick laid. The mason can, by bending over to pick up each brick and its mortar from the ground, lay 125 an hour.
- The contractor gains a profit of 0.2 cents per brick laid.

The customer therefore pays 0.8 cents per brick laid, while the mason earns 37.5 cents an hour or $3.00 for an eight-hour day. The contractor gets a profit of 25 cents per labor-hour. The customers believe they are paying too much, the masons believe they are getting far too little for back-breaking skilled labor, and the contractor thinks it is earning far too little profit. *All of them are right.*

Now suppose that Jack Cade, as portrayed by William Shakespeare in *King Henry VI* Part 2, decrees that the masons shall be paid 0.5 cents per brick laid to raise their pay to $5 for an eight-hour day. (This is comparable to what the Ford Motor Company would later pay for an eight- or nine-hour day.) Cade's actual decree was as follows, although it involved price controls rather than mandatory minimum wages.

> Be brave, then; for your captain is brave, and vows reformation. There shall be in England seven halfpenny loaves sold for a penny; the three-hooped pot shall have ten hoops; and I will make it felony to drink small [weak] beer. All the realm shall be in common; and in Cheapside shall my palfrey go to grass; and when I am king, as king I will be … there shall be no money; all shall eat and drink on my score, and I will apparel them all in one livery, that they may agree like brothers and worship me their lord.

Shakespeare wrote this comic relief speech to appeal to the groundlings, the lower-class English people who could not afford to pay for theater seats and therefore had to sit on the ground. While the groundlings were probably illiterate, they were not stupid. They understood that if bakers were told they had to sell seven halfpenny loaves for a penny, people would have to buy bread on the black market if they could get it at all.

Modern politicians seem to believe, however, that they can decree a $15 an hour minimum wage without consequences. If we return to the bricklaying example, the price per brick laid will increase from 0.8 cents to 1 cent which means there will be less demand for construction. The contractor will then have to cut the hours of its workers, or lay them off, which means they will be lucky to get the $3.00 a day they got previously. The $3.00 will also buy less because higher prices are inflationary.

Frank Gilbreth's introduction of a non-stooping scaffold delivered the bricks at waist level, which enabled the workers to lay 350 rather than 125 bricks per hour, and with less effort. Now we will try some new figures.

- A brick and its mortar still cost 0.3 cents.
- The mason is paid only 0.2 cents per brick laid but can now lay 350 an hour, which means he gets 70 cents per hour or $5.60 for an eight-hour day.
- The contractor accepts a profit of only 0.1 cents per brick laid, which however comes to 35 rather than 25 cents per labor-hour.
- The customer's price is now 0.6 rather than 0.8 cents per brick laid, which increases the demand for construction to keep the brick layers employed at the higher production rate and higher daily wage.

All three stakeholders or relevant interested parties, i.e., the customer, employer, and worker, are much better off under these conditions. The increased productivity not only supports the higher wage, it also drives down prices which means each dollar buys more rather than less. This underscores the fact, as stated by Henry Ford, that we cannot get something for nothing, but we can get something from what was believed to be nothing.

A couple of simple pictures will put the matter to rest very decisively. Organizations look for cheap labor because they believe that wages, prices, and profits are a zero-sum proposition in which one must suffer if another is to gain. Suppose we envision the price paid by the customer as a coin for which the customer expects to receive value and that the sole cost of the product or service consists of labor. The zero-sum assumption is that the customer pays for labor and profits, and the only way to get a lower price and/or more profit is to reduce wages as shown in Figure 4.4.

When we realize that the customer's actual costs include not just labor and the seller's profit, but also the waste that is built into the job, we come to a very different conclusion. The customer actually pays, and more than it should, not just for the seller's profit (less than the seller should get) and wages (lower than the workers should receive) but also waste as shown in Figure 4.5. *Each party to the transaction—seller, labor, and customer—perceives it is being shortchanged, and all are right.*

It is now obvious that the removal of waste allows the seller to increase its profits and pay higher wages and reduce the customer's cost in the bargain. Figure 4.5 also reiterates the previous observation that waste is inflationary because it increases prices but is, by definition, without utility or value. It's that simple.

We have seen so far, then, that science can deliver high wages side by side with high profits and low prices that legislation cannot. If however fewer workers can deliver the same output, will this put people out of jobs? The Luddites have made this argument for roughly 2,000 years, and they have been wrong for 2,000 years.

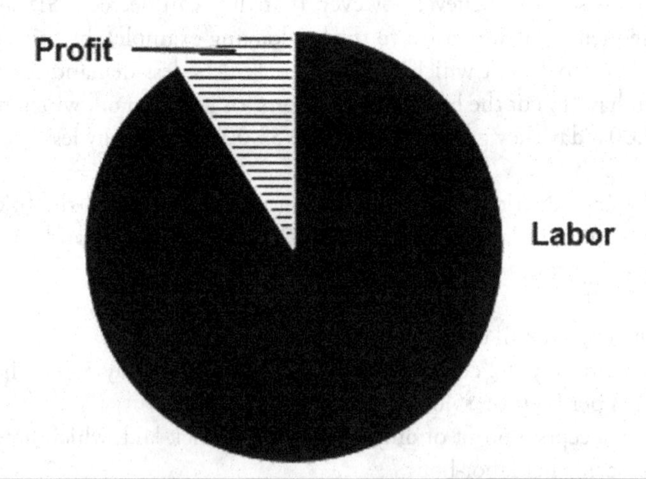

Figure 4.4 Perceived Customer Expenditure

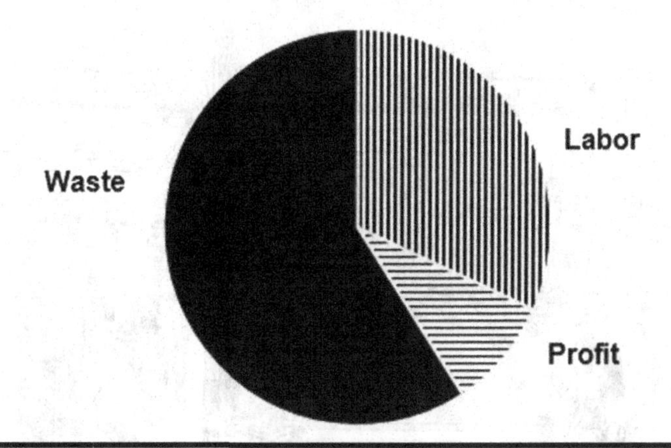

Figure 4.5 Actual Customer Expenditure

Lose the Luddites

Luddites believed that automation would put them out of work, so they sabotaged machinery during the late eighteenth and early nineteenth century as shown in Figure 4.6. Ford (1926, 157) later depicted Luddism as "the theory that there is only so much work in the world to do and it must be strung out." Luddism is therefore yet another manifestation of the dysfunctional scarcity mentality that says full employment for X equates to unemployment for Y and that union contracts and similar arrangements must limit productivity to protect jobs. Unintentional much less intentional limitations on productivity are almost certain to destroy jobs instead by inflating the cost of the goods or services in question. The abundance mentality says instead that higher productivity protects jobs by making the output less expensive while it also delivers higher wages and higher profits. This is a win-win outcome for all stakeholders.

Luddism, however, dates back hundreds if not thousands of years. James and Thorpe (1994, 388–389) describe how Roman Gaul developed the vallus, an animal-driven reaping machine that could harvest far more than scythe-wielding agricultural workers (Figure 4.7). The obvious drawback to an animal-drawn reaper was that the donkey or horse would trample the crops before they could be harvested, but the Romans got around this by having the animal push rather than pull the device. The use of vallus never became widespread, however, because, as stated by the reference, "the danger of social upheaval was always present and might have become acute if slaves started to be displaced by machines," as if that would have been a Bad Thing. The Romans did not share our modern views of basic human rights, but they did understand denarii, and they ought to have realized that one or two paid workers with a vallus (plus food and shelter for the horse or donkey) would have cost far fewer denarii than the upkeep of the necessary slaves.

The same reference says that China described a push scythe in 1313, but this was not adopted either because of bureaucratic opposition to labor-saving machinery; it would have put peasants out of work. Nobody ever considered the sensible idea that if the peasants could get more work done every day, the country would have become much wealthier. The mechanical reaper did not reappear until the early nineteenth century, when it was described in J.C. Loudon's *Encyclopedia of Agriculture* (1825 or 1835 edition). A Briton, John Ridley, who emigrated to Australia, developed a reaping machine to help overcome a labor shortage there. Pictures of the original Ridley stripper

Figure 4.6 Luddites smashing a textile machine, 1812 (public domain due to age)

Figure 4.7 The Vallus or Push Scythe. Loudon, John Claudius. 1883. (An Encyclopedia of Agriculture public domain due to age)

are almost identical to that of the vallus, although horses pulled a subsequent version from the front but with the reaper out of the path of the horse's hooves (Figure 4.8).

Luddism was hardly limited to uneducated or ignorant people. Queen Elizabeth I (Figure 4.9), an otherwise very capable monarch, refused to grant a patent for an automated knitting machine because she thought it would put knitters out of work (Conniff, 2011).

Crow (1943, 37) reports however that, by 1812, automation had made textile workers 200 times as productive as they were previously. This meant they could be paid much higher wages while the labor cost per yard of cloth would be negligible. Clothing prices could be lowered, and people could purchase more. This reference adds that, in 1844, *Merchants Magazine* reported that shirt cloth, which had cost 62 cents a yard 30 years previously, now cost 11–12 cents per yard. More could therefore be sold to keep the textile workers employed at higher wages. There was a time when many people could afford only one or two sets of clothing, but it is so relatively cheap today that most of us have closets full of clothing we never use.

Shoe Manufacture: A Case Study

The same principle, namely that lower prices increase demand to keep the workers employed at higher wages, carries over into shoe making as depicted by Taylor (1911) (emphasis is mine).

> The great majority of workmen still believe that if they were to work at their best speed they would be doing a great injustice to the whole trade by throwing a lot of men out of work, and yet the history of the development of each trade shows that each improvement, whether it be the invention of a new machine or the introduction of a

Figure 4.8 Nineteenth-Century Adaptation of the Vallus. Ridley stripper harvester in use on Canning Downs Station, Queensland, 1894 (public domain due to age)

Figure 4.9 A Famous Luddite. Queen Elizabeth I, "Darnley Portrait," circa 1575 (public domain due to age)

better method, which results in increasing the productive capacity of the men in the trade and cheapening the costs, instead of throwing men out of work make in the end work for more men.

The cheapening of any article in common use almost immediately results in a largely increased demand for that article. Take the case of shoes, for instance. The introduction of machinery for doing every element of the work which was formerly done by hand has resulted in making shoes at a fraction of their former labor cost, and in selling them so cheap that now almost every man, woman, and child in the working-classes buys one or two pairs of shoes per year, and wears shoes all the time, whereas formerly each workman bought perhaps one pair of shoes every five years, and went barefoot most of the time, wearing shoes only as a luxury or as a matter of the sternest necessity. *In spite of the enormously increased output of shoes per workman, which has come with shoe machinery, the demand for shoes has so increased that there are relatively more men working in the shoe industry now than ever before.*

Recall that Stern (1939, 16) reported that the labor-hours necessary to manufacture a pair of shoes had declined to less than an hour by 1936. This is why ordinary Americans now have closets full of shoes in contrast to the one pair they might have been lucky to own in the mid-nineteenth century. This leads to yet another question. Even if American hourly labor were to cost $40 an hour in wages, taxes, and benefits in this industry, why are our retail stores filled with "imported" (i.e., probably from the PRC or a similar venue) shoes with price tags well upward of $150?

Ford (Ford and Crowther, 1922) wrote essentially the same thing.

> When shoes were made by hand, only the very well-to-do could own more than a single pair of shoes, and most working people went barefooted in summer. Now, hardly any one has only one pair of shoes, and shoe making is a great industry.

He applied this principle to automobile manufacture by getting the price down to what most people could pay, which allowed him to employ far more autoworkers at high wages than had been employed previously when motor vehicles were luxuries for rich people.

The Ford Motor Company: Early Twenty-First Century

Shirouzu (2001) reported that Ford Vice President Jim Padilla "is trying to entice assembly workers and engineers to abandon nearly all they know about the mass-manufacturing system that Henry Ford brought to life about 90 years ago." It turns out that he was actually trying to reintroduce Ford's methods which, at the time, most American industrialists believed were Japanese in origin. Shirouzu, in fact, says lean manufacturing was developed at Toyota during the 1950s even though most of the techniques were developed by the Ford Motor Company during the first quarter of the twentieth century.

These techniques include just-in-time (JIT) production control systems or at least synchronized production systems that minimize cycle time and therefore inventory. The appearance of inventory at Henry Ford's plants was a visual indicator of a machine stoppage or similar trouble because parts were supposed to be used as rapidly as they were created. Standardization, the best-known way to do a job, was cited very explicitly by Frederick Winslow Taylor as well as Henry Ford.

Ford also wrote that no job should ever require a worker to take more than one step in any direction (as walking does not add value) or bend over (which, as shown by the Gilbreth bricklaying example, does not add value). A team leader wanted to change a work area to save another worker 2,000 steps per day. Other workers accused him of trying to help the company downsize, and the reference adds the entire lean initiative, "But it is one that some workers said they regard suspiciously because they worry that Mr. Padilla ultimately intends to increase efficiency and eliminate jobs." The truth is of course that people cannot be paid to walk, and if they must spend 20 minutes to do it—this is roughly how long it takes to walk slightly more than a mile—roughly four percent of their eight-hour shift consists of waste motion. To put this in perspective, four percent of a 250-day work-year comes to 10 days of waste that add no value for the employer, worker, or customer.

Self-Service Kiosks

A popular Internet meme features a picture of a self-service kiosk through which customers can place and pay for fast-food orders without involving a cashier. The caption says, "$15 an hour? Say hello to your replacement." The truth is however that if the customer must give the order to a cashier who must then give it to the food preparers, order placement essentially happens twice which is non-value-adding duplication of effort. If the job can be done through the kiosk, then the workers can focus on value-adding preparation of the actual food which creates the value for which the customer will pay.

There are also kiosks, or table-mounted screens, in some restaurants where customers can place and pay for their orders without involving the server. This means the server can deliver food to more tables and thus earn more pay per hour while the customers pay less per meal. This again creates a situation in which all the stakeholders (restaurant owners, customers, and servers) benefit at the expense of what was formerly waste.

Longshoremen versus Bar Code Scanners

"On the Pre-modern Waterfront" (2002, no author given) reports that the International Longshore and Warehouse Union (ILWU) resisted the introduction of bar code scanners even though the Pacific Maritime Association assured the union that nobody would lose his or her job due to automation. "But scanners don't pay dues, so the unions are balking at even that ancient innovation." Gross (2002) added of this, "Ever since the followers of Ned Ludd busted up British textile factories in the early 1800s, organized workers have resisted new technology." The union needed to recognize that if each longshore worker could get more done with a bar code scanner, he or she could receive higher wages. Gallagher (2002) reports however that the ILWU did eventually accept the scanners, and the workers received a base pay increase.

Luddism and Mechanical Power

Suppose for example that you are living on a paltry allowance that barely provides food, clothing, shelter, and other basic necessities of life, and you suddenly inherit two dozen mechanical servants. Would it make any sense to destroy that inheritance for fear that it might leave you unemployed? Emerson (1924, 352) put the matter as follows (emphasis is mine):

> We are like a young man until recently on scant allowance who has suddenly inherited an immense fortune. In the United States the uncarnate [inanimate, not made of flesh] energy used is thirty times as great as was the incarnate energy sixty years ago; it is as if each head of a family had inherited thirty slaves forced to labor for him without pay beyond the obligation to maintain. It is increasingly less the hard muscular labor of the hands and body that counts, it is more and more the intelligence to direct mechanical slaves that counts. *The man who smashes a machine because he fears it will take his job, the man who refuses the promotion due him for efficient control, misses the richest gift that any generation has ever been offered.*

"The man who refuses the promotion due him for efficient control" refers to those who resist efforts to make their jobs more efficient and productive and therefore a promotion from low-wage monotonous labor to high-wage intelligent craftsmanship. Managers and executives must ensure that the promotion is forthcoming, though, lest they prove the Luddites right.

Don't Prove the Luddites Right

Recall that, in 2001, Ford workers were suspicious of efficiency improvements because they were afraid this would result in layoffs. The company's history, including the labor relations problems during the late 1930s, offered some justification. Sinclair (1937, 81) wrote,

> Twenty men who had been making a certain part would see a new machine brought in and set up [at the River Rouge plant], and one of them would be taught to operate it and do the work of the twenty. The other nineteen wouldn't be fired right away—there appeared to be a rule against that. The foreman would put them at other work, and presently he would start to 'ride' them, and the men would know exactly what that meant.

This suggests that Ford himself had a no-layoff rule that said nobody would be discharged as a result of better efficiency, but his successors worked around that rule by looking for "causes" to fire the workers or "riding" them to make the work environment sufficiently hostile to get them to quit. The same kind of hourly workers who had, under Ford's personal governance, actually turned against their own union in the United Kingdom when asked to walk out on strike welcomed the United Auto Workers as a result of this direct violation of Ford's own principles.

Recall that, in the Gilbreth example, we found that we could cut the piece rate and still pay the workers higher wages because they could lay far more bricks per hour. The contractor could also reduce its profit per brick and still, due to the increased productivity, make more per labor-hour. If the piece rate and per-unit profit are held steady, then prices cannot be reduced which means there will be no increased demand with which to keep the workers employed at 350 rather than 125 bricks per hour.

If, on the other hand, the employer decides to be greedy and cuts the piece rate 64 percent so the workers' hourly wages will not change at all, and/or lays people off, then the consequences will be obvious as stated by Taylor (1911).

> after a workman has had the price per piece of the work he is doing lowered two or three times as a result of his having worked harder and increased his output, he is likely entirely to lose sight of his employer's side of the case and become imbued with a grim determination to have no more cuts if soldiering will prevent it.

It is meanwhile a well-known principle that a no-layoff policy is a prerequisite for a successful lean manufacturing initiative because the instant workers realize that the employer will discharge them when higher productivity makes them temporarily superfluous, they will cease to support the program. DeLuzio (2021) warned that organizations should expect a "train wreck" if they do not have a no-layoff policy in this context. This is simply a restatement of what Taylor said about piece rates in the context of layoffs.

The workers must therefore believe fully that a share of the benefits of the improvements will show up in their pay envelopes. They will, under these conditions, look everywhere for ways to make their jobs and those of their neighbors more productive. It is a basic principle of lean manufacturing that the frontline workers who actually do the jobs are usually in the best positions to identify wastes of motion, time, materials, and energy. Henry Ford credited his workforce with most of the shop floor improvements that drove his business to phenomenal success. Ford (Ford and Crowther, 1922) explained (emphasis is mine),

> *It ought to be the employer's ambition, as leader, to pay better wages than any similar line of business, and it ought to be the workman's ambition to make this possible.* Of course there are men in all shops who seem to believe that if they do their best, it will be only for the employer's benefit—and not at all for their own. It is a pity that such a feeling should exist. But it does exist and perhaps it has some justification. If an employer urges men to do their best, and the men learn after a while that their best does not

bring any reward, then they naturally drop back into "getting by." [This is essentially what Taylor wrote about soldiering, or marking time.] *But if they see the fruits of hard work in their pay envelope—proof that harder work means higher pay—then also they begin to learn that they are a part of the business, and that its success depends on them and their success depends on it.*

This is how the employer gains the commitment and buy-in of the workforce that makes lean manufacturing work to get prices down and wages and profits up. It also proves that the Luddites have always been wrong and will always be wrong unless management does something to prove them right.

Summary

This chapter has addressed the following issues that result in dysfunctional decisions to offshore jobs or pay workers the lowest wages as possible.

1. The concept of "high-priced men," as Frederick Winslow Taylor phrased it, dates back to the Biblical story of Gideon in which 300 confident and well-disciplined soldiers proved superior to 32,000 hesitant and poorly disciplined rabble. It is far easier from a logistics standpoint, and this carries into modern industrial considerations such as floor space and supervision, to support a few high-wage workers than a lot of cheap ones. The high-priced people also have an incentive to support standardization, standard work, and continuous improvement while the poorly paid ones do not.
2. What Emerson (1924) calls near common sense is a shortsighted focus on immediate results, including dysfunctional financial metrics that do not even reflect the economic reality on the shop floor. Supernal common sense takes the long-term view. It is why the businesses that grew up to serve prospectors during the California Gold Rush are still around today while few of the prospectors are even remembered.
3. A cost accounting system must not become a suicide pact. Reliance on traditional cost accounting metrics for managerial decision-making can be, and has been, utterly ruinous.
4. No number of slaves, serfs, or other unpaid workers, let alone cheap workers, are a match for motivated high-wage workers especially when automation is involved. Mechanical power is infinitely cheaper than human or even animal power, and the only question is how to deploy it to perform relatively complicated tasks. Engineers are getting better at this task all the time. Per Carr (1987, June 7), "Advanced countries with high labor costs can use mechanization to overcome competitors who have access to cheap labor." We have known this since the days of Frederick Winslow Taylor and even Aristotle, but organizations have nonetheless looked to cheap labor for competitive advantage.
5. Low wages are generally symptomatic of high prices and low profits. The workers believe they are underpaid, the customers feel like they are paying too much, and the employer complains about low profits. *All of them are right.*
6. The Luddites, who believe efficiency improvements and automation will destroy jobs, have been wrong ever since there have been Luddites, and they will always be wrong unless the employer does something to prove them right. The employer must not discharge workers when productivity improvements allow fewer people to do the same amount of work, or workers will withhold their support for further improvements. Luddism otherwise creates unemployment by building waste into the prices of goods and services.

We cannot legislate high wages and/or low prices, but we can engineer them through productivity improvements. The next chapter will go into the details and discuss methods whose effectiveness was proven more than a hundred years ago. Nothing has happened to change the underlying principles, although modern technology provides opportunities that pioneers like Ford, Taylor, Edison, Gilbreth, and Emerson could only imagine.

Notes

1. The workforce was predominantly male in 1911; we would say "high-priced people" or "high-wage labor" today.
2. Emerson wrote this long before Las Vegas, which began to grow during the construction of the Hoover Dam and then added tourism to the state's income, reviving Nevada's fortunes.
3. No part of this book constitutes formal engineering advice, but the webinar is available periodically from Pennsylvania Training for Health and Safety for anybody who wants extensive details on occupational health and safety aspects of hexavalent chromium compounds. The intended takeaway here is that everything the employer has to do to protect employees from these materials will add costs that do not appear in the obvious places, and this principle carries over into other hazardous materials as well.

Chapter 5

We Can Do It!

The Introduction stated that reshoring is a Specific, Measurable, Achievable, Realistic, and Timely (SMART) goal. The book showed subsequently that 100-fold or even greater productivity improvements are not only achievable and realistic but are matters of historical record. Emerson (1924, 27–28) wrote of the gap between existing and achievable performance as follows.

> American organization for operation … proves on investigation to be inefficient, often disgracefully so, the efficiency of the output of men of militia age of the country as a whole being not more than 5 per cent, the efficiency of use of materials being not more than 60 per cent, the efficiency of equipment facilities not averaging 30 per cent. These inefficiency statements can be verified from the facts, by any competent experts, as readily as an assayer.

Recall also that Henry Ford (Ford and Crowther, 1922), who grew up on a farm, also cited the five percent figure for labor efficiency, at least in agriculture. "The farmer makes too complex an affair out of his daily work. I believe that the average farmer puts to a really useful purpose only about 5 per cent of the energy that he spends." He included examples such as carrying buckets of water rather than conveying it by pipes or conduits, a practice that was known even in ancient times. The counterweight water lift known as the shaduf dates back purportedly to 3000 BCE and was used extensively in ancient Egypt, and Archimedes' screw dates back to the time of Archimedes.

The bottom line is therefore not only that "we can do it" but that we actually did it. We did it in the first part of the twentieth century thanks to people like Henry Ford, Frederick Winslow Taylor, and Frank Gilbreth. We did it between 1941 and 1945, as shown in Figure 5.1, when the civilized world was fighting for its life against Axis tyranny.

We did not import weapons and weapon components from what was then China; we made these things and used them to arm the Chinese, Russians, British, and other allied nations. The Murmansk Run, for example, delivered American-made war materials to Russia for use against the Nazis on the Eastern Front. Nothing but self-limiting thinking prevents us from doing it again today to which Hanson (2020) adds, "we can wake up as we did on Dec. 8, 1941, to ensure that Americans control their own fundamentals of life—food, fuel, medicine and strategic industries—without dependency on illiberal regimes." This book has shown so far that:

DOI: 10.4324/9781003372677-5

Figure 5.1 Women Shipfitters, Second World War. US Department of the Navy, 1943. "Women shipfitters worked on board the USS NEREUS, and are shown as they neared completion of the floor in a part of the engine room. Left to right are Shipfitters Betty Pierce, Lola Thomas, Margaret Houston Thelma Mort and Katie Stanfill. US Navy Yard, Mare Island, CA." (Public domain as a US Government photograph)

1. Waste (muda) is inflationary, and so are higher wages that are not commensurate with higher productivity. Higher productivity, on the other hand, enables higher wages side by side with lower prices. It also generates taxable economic activity with which to reverse deficit spending.
2. The decline or absence of manufacturing capability is a *universal* leading indicator of national decline.
3. The People's Republic of China (PRC) is a dangerous geopolitical rival that sells counterfeit and substandard products to American consumers and businesses and has recently threatened to use its control of vital supply chains to harm the United States. International supply chains are meanwhile vulnerable to force majeure even when trustworthy trading partners are involved.
4. The perceived need for cheap labor is a dysfunctional paradigm that, among other things, gives enormous quantities of waste a place to hide while it introduces major supply chain risks.

The next question is what to do about this, and the book has already discussed some of the answers. Recall for example that Frank Gilbreth proved that bricklaying, as practiced throughout most of human history, wasted 64 percent of the workers' labor on waste motion. Then again, this chapter will show that Gilbreth may have reinvented methods that were known in medieval times but forgotten. The lesson from this and similar examples is that we need to pay close attention to everything we see because success secrets often hide in plain view. If we paraphrase Sir Arthur Conan Doyle's Sherlock Holmes, "many see but few observe." It does not help to go to gemba (the value-adding workplace) and watch what happens if we don't know what to look for. This chapter will hopefully equip the reader to do so.

Aristotle predicted, and this was when machines such as industrial robots existed only in Greek legends, that automation would render slavery and low-wage labor uneconomical. Thomas Edison (Benson, 1923, 46) predicted that sufficient automation would eliminate the need for even medium-wage labor because once machines could do everything, "Manufacturers can then afford to pay any kind of wages." This leads to the basic principle that *if we make jobs sufficiently productive, their per-unit labor cost is negligible regardless of the worker's hourly wage.*

Emerson (1909, 16) added this,

> It is distinctly the business of the engineer to lessen waste wastes of material, wastes of friction, wastes of design, wastes of effort, wastes due to crude organization and administration in a word, wastes due to inefficiency. The field is the largest and richest to which any worker was ever turned … The field is large and rich because so little is being done, because there is so much to do.

When people such as Henry Ford and his contemporaries began to do it, the United States started on an upward trajectory to unprecedented wealth, prosperity, and national power that ended only when short-sighted executives, the kind of people whom Ford said should never run a business because their eyes were always on the dollar rather than the job that produced the dollar, began to send jobs offshore for cheap labor. This is the business of the reader and the decision-maker, it ought to be the business of the nation's lawmakers, and this chapter will show how to make this happen.

Think Like a Greek

We don't need a new way of thinking, but rather the thought process we inherited from the ancient Greeks. While all societies of that era had myths and legends about gods, demigods, and heroes, the Greeks were, as far as I know, the only people to write what we might now call science fiction stories. The difference between myths and legends, and science fiction, is that the former rely on forces that are available only to gods and other superhuman entities while the latter rely on their achievability by ordinary humans.

The Greek stories included flight (Daedalus and Icarus), robots like the servants of Hephaestus, and even the mecha (combat robots) that were made popular by Japanese anime films. "Johnny Sokko and His Flying Robot" features, for example, what looks like a giant robotic samurai who differs little (other than for his ability to fly and use modern weapons) from the bronze giant Talos, whom Hephaestus built to defend the island of Crete. While Hephaestus's robots were divine creations, Greek inventors like Heron of Alexandria created similar albeit far lesser mechanical servants such as automatic doors. Other societies portrayed mortals at the essential mercy of their gods although some, like the Hindus, made it clear that even the gods were subject to laws like Dharma (the Right Way). Greek stories, on the other hand, showed that mortals could fight and even defeat gods, as Diomedes did when he wounded Ares during the Trojan War.

The Greeks created their gods in their own images, as depicted by Aristotle (played by Barry Jones) in the movie Alexander the Great (1956) starring Richard Burton. "The gods of the Greeks are made in the image of Men …who can be understood and felt." These gods had human faults such as Zeus's infamous infidelity toward his wife (hence all the demigods such as Hercules, Perseus, and so on) and Ares's violent recklessness; he was a good role model for how a soldier should not behave. They were also, however, role models to whose achievements mortals could aspire. Hephaestus invented an enormous array of mechanical devices, and Athena taught that cleverness (e.g., the Trojan Horse) was often more important than brute force. When the giant brothers Otus and Ephialtes waged war on the gods, Ares rushed out headlong to fight them and was quickly stuffed into a jar. One version of the story adds that they could not be harmed except by one another, so Artemis transformed herself into a deer and ran between them while they were hunting. They hurled their javelins at her, but they didn't follow a basic safety rule, namely to always know what is behind one's target, so they died unwittingly at one another's hands. Stories of this nature doubtlessly taught the Greeks how to outthink and thus defeat stronger opponents.

The concept of humans as the arbiters of their own fates, rather than subjects of the whims of gods and even natural forces, is meanwhile reinforced by another statement by Barry Jones as Aristotle: "Wonders are many, but none is more wonderful than man himself." A search on this quote shows, however, that it was not originally from Aristotle but rather from Sophocles' *Antigone* (as translated by R.C. Jebb, 1893).

> Wonders are many, and none is more wonderful than man; the power that crosses the white sea, driven by the stormy south-wind, making a path under surges that threaten to engulf him; and Earth, the eldest of the gods, the immortal, the unwearied, doth he wear, turning the soil with the offspring of horses, as the ploughs go to and fro from year to year.
>
> And the light-hearted race of birds, and the tribes of savage beasts, and the sea-brood of the deep, he snares in the meshes of his woven toils, he leads captive, man excellent in wit. And he masters by his arts the beast whose lair is in the wilds, who roams the hills; he tames the horse of shaggy mane, he puts the yoke upon its neck, he tames the tireless mountain bull.
>
> And speech, and wind-swift thought, and all the moods that mould a state, hath he taught himself; and how to flee the arrows of the frost, when 'tis hard lodging under the clear sky, and the arrows of the rushing rain; yea, he hath resource for all; without resource he meets nothing that must come: only against Death shall he call for aid in vain; but from baffling maladies he hath devised escapes.

Sophocles therefore conveys the lesson that everything that humanity possesses is the creation of humans rather than gods. Agriculture, domestication of far stronger animals, protection from the weather, remedies for "baffling maladies," and even speech and thought are all human rather than divine creations. This kind of empowering language appears to have made the ancient Greeks more willing than their contemporaries to challenge the proposition that "this is the way things are" means "this is how it must always be."

Another story about Daedalus teaches meanwhile the lesson that very simple solutions often hide in plain view. Somebody challenged Daedalus to thread a sea shell, which was a seemingly impossible task. He completed it by attaching the thread to an ant, putting the ant into the shell, and putting some honey at the tiny opening at the other end to attract the insect.

Stories of this nature have powerful influences on human behavior because people who read them think, "Can I actually go out and do this?" The stories of non-Greek societies certainly included role models for ideal human behavior, but few if any of them suggested that humans might fly or build mechanical servants without the aid of sorcery or divine aid. The Greeks did try to make their stories into reality, and they enjoyed enormous success despite the almost complete absence of modern science. Archimedes built mechanical devices that could purportedly lift enemy warships out of the water or crush them, along with a heat ray that used reflected solar energy to threaten them with fire; the mortal enemy of any sailor in a wooden warship. The actual effectiveness of the heat ray is still open to debate but today's solar furnaces rely upon an idea that is more than 2,000 years old.

Learning from Hercules

The Labors of Hercules, a set of twelve tasks that he had to perform, show meanwhile that the hero often had to think around problems rather than use his superhuman strength to resolve them, just as modern plant managers who want to double their output often don't need to double their workforces or plant and equipment. Henry Ford (Ford and Crowther, 1922) wrote of his era's farmers, "His whole idea, when there is extra work to do, is to hire extra men." This is the kind of thinking that encourages executives to go offshore for cheap labor rather than think around the problem at hand. Hercules didn't need bigger muscles to solve most of his problems; he needed only to "put more brains into his business—to overcome by management what other people try to overcome by wage reduction" as stated by Ford.

One of Hercules's tasks was to kill the Nemean Lion, whose hide was similar to modern Kevlar as it was impenetrable by arrows and spears. This ruled out the traditional way to kill a lion, but Hercules realized that the animal's skin had to be flexible to allow it to move, and he strangled it accordingly. He also discovered that the lion's claws could cut its skin, so he took its impenetrable hide for himself. This teaches that one can learn from almost anybody or anything and that one must avail oneself of whatever opportunity comes to hand. Hercules did not think, "The lion is dead, my Labor is complete." He realized, "this lion skin is better armor than anything I can get from the best bronze-smith in Greece."

Another task that was intended to demean Hercules was to compel him to clean out the Augean Stables, but he did not even consider the use of the traditional shovel for this task. He diverted two rivers to flush the stables clean in a single day. *The basic principle is to never assume that the obvious way to do a job is the best way, even if people have been doing it that way for centuries.*

This legacy from Greek history carries over into modern entertainments such as the MacGyver television series, in which the hero often had to improvise solutions from whatever was available nearby. The show's popularity even created the verb "to MacGyver something," which means to come up with an innovative solution from whatever is available. The corresponding French word bricolage means essentially the same thing. The support team for the Apollo 13 mission had to do exactly that with items that were available on the spaceship to allow the crew to effect lifesaving emergency repairs.

There is meanwhile a stereotype to the effect that Asian cultures are not inventive or at least have not been inventive in the past. They were generally not familiar with Greek legends, although some of these may have passed into India via Alexander the Great. The *Tao Te Ching*, while an otherwise outstanding book on leadership and governance, advises the ruler to keep his or her people's bellies full and their minds empty, and their bodies strong but their minds weak (Lao Tzu, 1963, 59).

This supports a very harmonious society, especially so if people have no desire to rise above their stations in life, but societies that think this way do not invent flying machines and robots. When Tokugawa Ieyasu ("Yoshi Toranaga" in James Clavell's *Shogun*) gained control of Japan, he largely closed the nation to the outside world to prevent importation of ideas that might upset Japan's traditions and social order. This worked, at least until Admiral Perry showed up in the nineteenth century with steam-powered cannon-armed ships while the Shogunate's few firearm-equipped soldiers still wielded arquebus muskets of the sixteenth-century design. The Japanese realized that they had to adapt quickly, and they imported European science and even, as depicted previously, adapted Prussian military principles of organization to industrial management.

Learn from Everything You Encounter

This is not to say that non-Greek societies did not invent anything; James and Thorpe's (1994) *Ancient Inventions* depicts some very advanced technology from all over the pre-modern world. While Europeans wrote on wax or clay tablets, and later parchment (sheep skin), China invented paper. The Egyptians built pyramids and even a precursor to the Suez Canal. The Aztecs, who were essentially Stone Age people, built very productive floating farms known as chinampas. Ebel (2019) shows that chinampas could be quite viable in the twenty-first century; "The chinampa system is still practiced in suburban and inner city agriculture (Leon-Porfilla, 1992). It is one of the most intensive and productive production systems ever developed (Altieri and Koohafkan, 2004), and it is highly sustainable." The latter reference elaborates that the soil in the chinampas is enriched by organic matter from aquatic plants plus small quantities of animal manure, and the animals can eat otherwise unusable produce from the chinampas. The concept appears similar to that of Upward Farms' (upwardfarms.com) production of striped bass whose waste fertilizes crops, a portion of which is eaten by the bass. This chapter discusses Upward Farms' approach in more detail later.

An obvious advantage, which is shared with vertical farms in urban environments, is that vegetables grown in the chinampas do not require much transportation (a non-value-adding activity) to get them to customers. This underscores the vital takeaway that we need to pay attention to everything we observe. The fact that something is hundreds or even thousands of years old does not mean it will not work, especially with improvements in modern technology.

The Aztecs used cotton armor known as ichcahuipilli. While "cotton" and "armor" sound like contradictory terms, it is to be remembered that modern Kevlar is similarly made from a non-metallic (aramid) fiber and works by dissipating a projectile's impact rather than trying to stop it outright. The Aztec version was so effective, and also so comfortable in hot weather, that Spanish Conquistadors favored it over European plate armor. Wikipedia cites Phillips and Jones (2015) to claim that inch-thick cotton armor could stop arrows and even musket bullets, which could have conceivably made it useful on European battlefields.

The Greeks meanwhile had the linothorax or breastplate made from layers of linen rather than bronze. Moore (2011) reports that a replica of this kind of armor, which is lighter than bronze or steel, would stop an arrow or a sword cut. James and Thorpe (1994, 211) add that China developed paper armor. We don't think of paper as capable of stopping anything, but layered and pleated paper could stop arrows. The principle seems again to be to dissipate the projectile's impact rather than to try to stop it outright.

Casimir Zeglen meanwhile developed armor made from properly woven silk of which an inch's thickness would stop most bullets. This armor weighed only two pounds per square foot. Addition of a tenth-inch-thick steel armor more than doubled the weight but would stop steel-jacketed rifle bullets (Simon, 2005). Had Archduke Francis Ferdinand worn his silk vest on June 28, 1914, and had it extended to cover his neck, the First World War would have probably not happened. This again reinforces the lesson that we need to learn from everything we encounter, and research is now being done as to whether spider silk might be superior to Kevlar for body armor.

If we return again to the Greeks, the potential invention of the railroad more than 2,000 years ago underscores the need to pay attention to everything. The Greeks did build the Diolkos, a paved trackway for overland transportation of ships to avoid the need to circumnavigate the Peloponnese, which was a dangerous voyage at the time. James and Thorpe (1994, 133–135) add that the Greeks also used pistons, cylinders, and valves as exemplified by a water pump for putting out fires. These, along with Heron of Alexandria's demonstration of the ability of steam to do useful work, would have provided all the basic technology necessary for a railroad engine. As matters stood, the idea of a railway was forgotten until medieval times when wagonways, which were essentially railroad tracks for horse-drawn carts, were used in mining and other occupations.

Now, however, Asian countries have the same access to Greek stories that the West has to classics such as Lao Tzu's *Tao Te Ching*, Sun Tzu's *Art of War*, and the Hindu epics *Ramayana* and *Mahabharata*, and they have become far more innovative as a result. Shigeo Shingo, for example, deployed very simple and low-cost error-proofing solutions to eliminate a very wide variety of defects and nonconformances. Westerners can meanwhile learn much from Asian classics on governance, and the exchange of ideas will doubtlessly benefit everybody involved.

Break Paradigms and Think Around Problems

We have seen so far that the legendary hero Hercules had to step outside of paradigms or self-limiting ways of thinking. An invulnerable lion did not have to be killed with spears or arrows, and enormous piles of animal waste did not have to be shoveled. Water power can be used instead to perform literally Herculean tasks. Hercules's legacy underscores further the power of the Greek thought process to change the entire world.

While Hercules might have been a legendary figure, Alexander III of Macedon was very real, and his legacy persists to this very day. He opened commerce and an exchange of ideas between Europe and India, and his influence still persists in Central Asia where Kandahar in Afghanistan is named for him. Alexander cited Hercules as a role model, wore a lion skin to imitate him, and even thought like him. When he faced the challenge of the Gordian Knot, he realized that nowhere did the rules require him to untie it so he used his sword to cut it instead (Figure 5.2). The island fortress of Tyre was meanwhile seemingly impregnable but Alexander, who might have remembered Hercules's feat with the Augean Stables, built a causeway to connect the fortress to the mainland. The causeway still exists today, and it supports many buildings. The Greek thought process, which includes specifically the ability and inclination to "think outside the box" and to "MacGyver solutions" literally conquered the known world long ago. Its application today can conquer seemingly insurmountable problems that encourage short-sighted planners to offshore valuable jobs that belong in the United States.

There was meanwhile one Greek who could have out-Alexandered Alexander himself, and that was Memnon of Rhodes.[1] Memnon realized that the Persians who had hired him could not defeat Alexander on the battlefield, so he advised them to retreat while devastating everything behind them to deny the Macedonian Army the ability to live off the land. This "scorched earth" strategy was later made famous by Russia, but the Persians were unwilling to retreat from the Macedonian "barbarians," which is why so many cities in that region are now named "Alexandria." These include Mashhad (Alexandria Susia) in Iran.

Figure 5.2 Alexander Cuts the Gordian Knot. (By André Castaigne, public domain due to age)

Our Legacy from Alexander the Great (and Henry Ford)

Here is a legacy from Alexander that anybody can take into any workplace, and it tells us that we do not have to be satisfied with the way the job is currently designed or that the job cannot be improved; "There is nothing impossible to him who will try." These words are the sword that will cut any modern Gordian Knot that consists of preconceived ideas and long-standing habits whose sole effect is to retard progress. Recall that Henry Ford (1926, 2) wrote similarly more than 2,000 years later, "Only the old, outworn notions stand in the way of these new ideas. The world shackles itself, blinds its eyes, and then wonders why it cannot run!" This chapter will cite examples including the self-limiting paradigms that meat must come from animals and that farms must operate as they have for thousands of years. Modern technology now offers ways to eliminate enormous amounts of waste that go with both models for food production.

When Education Is Dangerous

Education is normally valuable because it teaches people what others have done successfully in the past, but it can be dangerous if it teaches people that something cannot be done. Ford (Ford and Crowther, 1922) wrote of this,

> All the wise people demonstrated conclusively that the [internal combustion] engine could not compete with steam. They never thought that it might carve out a career for itself. That is the way with wise people—they are so wise and practical that they always know to a dot just why something cannot be done; they always know the limitations. That is why I never employ an expert in full bloom. If ever I wanted to kill opposition by unfair means I would endow the opposition with experts. They would have so much good advice that I could be sure they would do little work.

Consider for example the matchsticks and dice production control exercise in Goldratt's and Cox's (1992) *The Goal*. Each workstation has a capacity of one to six as determined by a die roll. The workstation can process the lesser of the die roll and the incoming work, which means high die rolls for which no work is available are wasted. This teaches the valuable lesson that time lost at a constraint or capacity-constraining resource, and each workstation is a constraint because its capacity equals those of the others, is lost forever. This is why favorable variation does not offset unfavorable variation. The exercise also teaches ostensibly that it is impossible to run a balanced factory at full capacity. The reader, or student in a Theory of Constraints class, might go away with the wrong lesson, namely that a balanced factory cannot operate at full capacity without accumulating endless quantities of inventory as happens in the simulation.

The problem with this lesson is that Henry Ford did exactly that (Ford and Crowther, 1922): "The idea is that a man must not be hurried in his work—he must have every second necessary but not a single unnecessary second." The only way he could have achieved this was to have removed all the variation in processing and material transfer times from his operations, and there are abundant examples of how he did this. Operator-induced variation was removed through what we now call standard work, along with automation to remove the human element even further. Ford's operations have been compared to a clockwork mechanism whose inspiration came from the mechanical watches Ford repaired when he was a boy.

If one takes away from *The Goal* the idea that it is impossible to run a balanced factory at full capacity, then this knowledge is likely to work against the student by telling him or her that "it can't be done." If however the student takes away the lesson that variation in processing and material transfer times is an enemy of productivity that should be suppressed with whatever methods are at hand (such as standard work), then the knowledge empowers him or her.

Another example consists of the well-known fact that one can recover only a limited amount of energy from a generation process that converts heat into mechanical energy, as happens in a coal- or gas-fired power plant or an internal combustion engine. The theoretical limit is in fact set by the thermodynamic Carnot cycle, and real-world processes like the Rankine (steam) and Otto (internal combustion) cycles are even less efficient. If the only thing we learn from the Carnot cycle is, "We can't recover even fifty percent of the energy from gasoline, diesel fuel, coal, oil, or natural gas as useful mechanical or electrical power," this is a self-imposed limit that says "We can't." If however we realize, "We must bypass the Carnot cycle by converting the fuel directly into electrical power," this opens the door to solutions.

This is the same lesson we learned from the Gordian Knot. The assumption that the knot had to be untied, and it was designed to make this impossible, tells us, "We can't." If we ask, "How can we undo this knot without untying it?" we can solve the problem. The same applies if we ask, "How can we bypass the Carnot cycle and its real-world counterparts?" There is something called an internal reforming fuel cell that combines high-temperature steam with a hydrocarbon to generate hydrogen for use in fuel cells—and fuel cells are not subject to the Carnot cycle's limitations. The US Department of Energy (no date given) says "A conventional combustion-based power plant typically generates electricity at efficiencies of 33 to 35%, while fuel cell systems can generate electricity at efficiencies up to 60% (and even higher with cogeneration)." This chapter will show, in fact, that Emerson (1909, 17) pointed out long ago that fireflies convert chemical energy directly into light, which bypassed all the efficiency limits on the incandescent lights of his day. The firefly didn't tell us how to do this, but it proved it could be done.

We have also seen how education in traditional cost accounting can be dangerous if the accountant tries to apply the material to managerial decisions rather than to just the calculations required by the Internal Revenue Service and Securities and Exchange Commission. While accounting statements might include overhead and allocate it to specific products, for example, it is essentially a sunk cost that is not affected by the decision to make or not make products. When one understands the difference between cost accounting and tax accounting, and managerial or engineering economics that consider marginal revenues, costs, and profits instead, the education becomes useful.

The Basic Principles

We will begin by summarizing the basic principles that support reshoring of American jobs, along with enormous increases in productivity, wages, and profits from jobs that are already in the United States. The fundamental idea as pointed out by Edison is that *efficiency makes the per-unit labor cost negligible regardless of how well the worker might be paid.* We achieve efficiency through understanding of (1) supernal common sense, (2) friction, (3) opportunity costs, (4) power of observation, and (5) never assuming there is not a better way to do something. Ford (Ford and Crowther, 1922) wrote of the latter, "If there is any fixed theory—any fixed rule—it is that no job is being done well enough."

1. Harrington Emerson's *supernal common sense* means a focus on long-term rather than immediate results, the latter of which is the focus of near common sense. Supernal common sense includes assessment of the total cost of ownership (TCO) or total cost of use of a product or service as opposed to its price tag, and the latter is often an incentive for dysfunctional purchasing decisions. This includes the purchase of quantities unnecessary for immediate needs, acceptance of long and therefore risky supply chains in exchange for low prices, and long lead times for delivery of goods of questionable quality. The abundance mentality as depicted by Dr. Stephen Covey is another element of supernal common sense while the scarcity mentality is an element of near common sense. The scarcity mentality and near common sense suggest, for example, that removal of waste from jobs will displace workers while the

abundance mentality and supernal common sense tell us accurately that waste destroys jobs instead by pricing their outputs out of the market and delivering low wages in the bargain.

2. *Friction* is General Carl von Clausewitz's (Figure 5.3) (1976, 119) term for "the force that makes the apparently easy so difficult ... countless minor incidents—the kind you can never really foresee—combine to lower the general level of performance, so that one always falls short of the intended goal." It relates to seemingly minor inefficiencies whose cumulative effect is to undermine organizational performance, and perhaps fatally. Industrialists such as Henry Ford, Shigeo Shingo, and Taiichi Ohno later applied the same principle to manufacturing. Frank Gilbreth, Frederick Winslow Taylor, and others proved that improvements on the order of 300 percent are achievable through the elimination of friction. One exception appears to be a Japanese company that achieved a 100-fold improvement in the output of disposable gowns with items from a 100-yen (roughly one dollar) store (Toyota Times Global, 2020). This did not, unlike most breakthrough improvements of this nature, require

Figure 5.3 Carl von Clausewitz. Wach, Karl Wilhelm. (Portrait of Carl von Clausewitz, public domain due to age)

substantial capital investments in machinery. *Standard work* meanwhile removes variation, another form of friction, from the manner in which a job is performed.

3. *Opportunity costs* reflect the unrealized benefits, which are totally invisible to the cost accounting system because we cannot write them off as losses for financial and tax reporting purposes, of failure to adopt superior technology or equipment. This issue is yet another example of supernal versus near common sense. Attention to this issue is the source of 100-fold improvements but, with few exceptions such as the one cited above, substantial capital outlays may be necessary.
4. *Many see, few observe* or, as Sherlock Holmes put it, "You see but you do not observe" means that numerous opportunities exist in plain sight but are overlooked because people are used to things as they are and have no desire to improve them. As but one example, tens if not hundreds of millions of soldiers during the horse and musket era of warfare learned motion-efficient drills for loading muskets and cannons, but only in the early twentieth century did Frank Gilbreth recommend the application of the same principles to civilian jobs. The Ohno Circle, or a circle in which an observer stands as recommended by Taiichi Ohno to see what actually happens on the shop floor, is useful only if the observer does not take everything he or she sees for granted.
5. *There is nothing impossible to him who will try*. This is Alexander the Great's answer to what Shigeo Shingo called "nyet engineers," with "nyet" being Russian for "no," i.e., people who would offer reasons why something cannot be done. The famous Russian commander Alexander V. Suvorov (1729–1800) called these people *nichtwissers* (know-nothings), those who don't know and won't bother to find out. Field Marshal Helmuth von Moltke called them "men of the negative," those who were always ready to tell you why something wouldn't work but never offered constructive ideas of their own (Levinson, 2017). The term "concrete heads" (Fast, 2017) also has been used. "These are so-called leaders who don't get 'it,' don't want to get it and won't listen to those who do get it." The "concrete heads" not only won't do the job themselves; they won't let anybody else do it either. "Anchor-dragger" is yet another variant on nyet engineers, nichtwissers, and people of the negative.

It is meanwhile necessary to educate the American consumer to buy value and not waste. Offshoring will become far less attractive when consumers refuse to tolerate price tags suitable for American-made goods on poor-quality products from low-wage offshore labor. The fact that the seller puts a fancy brand name or a celebrity endorsement on these products does not make them worth any more money, and purchasers have the right to demand their money's worth.

Efficiency Makes the Per-Unit Labor Cost Negligible

Emerson (1909, 19) identified very clearly (1) the principle that most jobs contain enormous amounts of waste and (2) the need to share the gains of efficiency improvements with all stakeholders including especially the workers (emphasis is mine).

> That men should work very hard for 9 or 10 hours per day is not a hardship if they are interested in their work, or if, in the larger interest of the community, they work efficiently; but to work desperately hard for many hours at dirty, hot and rough work, yet waste 67 per cent of the time and effort, is unpardonable. What could have resulted from an elimination of this waste?

1. The product could have been cheapened.
2. The men could have worked one-third the time and have accomplished as much.
3. One man could have done all the work and have earned three times as much.

The benefits should however be distributed in all three directions. *Fewer men should work less hard, receive higher wages, and deliver a cheaper product.*

Consider for example Henry Ford's unprecedented $5 a day (62.5 cents per hour) minimum wage, which was a substantial amount of money in the first quarter of the twentieth century. It once took 12.5 work-hours, which would have cost $7.81 under the $5 a day minimum wage, to assemble a Model T. This was subsequently reduced to 93 minutes ("Moving assembly line debuts at Ford factory," no author given) which means the labor cost of an entire automobile was slightly less than a dollar in the money of the early 1920s. The Model T was priced at $850 which means the labor component of the vehicles was a little more than 0.1 percent of its total cost to the buyer. Most of the price consisted of the raw materials (such as coal and iron ore along with rubber and wood) that went into the vehicle along with the capital cost of Ford's manufacturing equipment, steel mill, and logistics system. The ratio has increased during the past ninety or so years; Morgan (2008) reports that about 10 percent of the cost of a new automobile consisted of labor in 2008. The automated nature of the manufacturing process means nonetheless that labor is not the principal cost contributor.

There is meanwhile no way that cheap offshore labor, or even offshore slave labor equipped with Ford's most advanced production methods, could compete with this. Even if the PRC were to adopt the most advanced lean manufacturing methods, it could not compete in the United States because the cost of moving its goods across the Pacific Ocean, along with inventory carrying costs and even the risk of the products falling overboard (Koh, Ann, 2021), would exceed the labor cost difference. Remember that the total cost of ownership consists of far more than the item's actual price, and it includes but is not limited to supply chain risks, lead times, and the cost of poor quality.

Friction and Opportunity Costs

Friction and opportunity costs are related but not identical. A good general rule for distinguishing between them is that the changes necessary to remove friction range from zero cost (such as rearranging a workplace to reduce the need to walk) to low cost, while removal of opportunity costs is likely to require moderate to substantial capital investments. This rule is however general rather than absolute because, as but one example, removal of variation (a form of friction) from processing and material transfer times may require retooling of batch operations to continual flow operations.

The difference between friction and opportunity costs can be illustrated with Frederick Winslow Taylor's improvement of pig iron handling at Bethlehem Steel in the late nineteenth century. Taylor removed friction from the job by having the worker pace himself, which improved productivity almost fourfold. Attention to the issue of whether workers should have been carrying pig iron at all, and the cost of the foregone opportunity to use mechanical conveyors instead, is the kind of thinking that delivers improvements of 100-fold and even more.

Another example of the removal of friction was Taylor's instruction to select the right shovel for the job. A mid-sized shovel had too much capacity for heavy iron ore, and too little for ash and similar materials, so the use of a small shovel for the former and a big one for the latter enabled the worker to achieve more with less effort. The concept of opportunity costs would have asked whether manual labor should have been used at all, or whether some kind of machine could supply the brute force instead. When the technology and/or capital for automation are not available, however, attention to the issue of friction can nonetheless deliver substantial improvements that enable higher wages, higher profits, and lower prices simultaneously.

Recall that friction relates to seemingly minor inefficiencies that degrade performance. Opportunity costs represent traditionally the income not realized from, for example, alternative A if one selects alternative B. Alternative B can include, however, the decision to do nothing. This can include failure to purchase a machine that can do the work of a dozen or even a thousand manual laborers.

Friction is a good catch-all term for inefficiency that is built into a particular job and is easily removable on the shop floor or equivalent. Edward Mott Woolley described a fabric folding operation in a bleaching and dyeing factory as follows:

> But all [employees] took two steps to the right to secure their cloth, returned to the tables, folded the stuff and deposited it on another pile two steps to the left. That had always been the practice; no one had ever thought to question it.
>
> (The System Company, 1911, 41)

A simple rearrangement of the workplace to eliminate the need to walk doubled the workers' productivity and probably reduced their exertion in the bargain, and the cost was essentially zero as nothing had to be purchased to make this work. Note however the phrase, "no one had ever thought to question it," a phrase that also reinforces the phrase "Many see, but few observe." Employees should define as "friction" any action that does not add value to the job.

Ford (Ford and Crowther, 1922) added, "The undirected worker spends more of his time walking about for materials and tools than he does in working; he gets small pay because pedestrianism is not a highly paid line." The need to search for tools and parts and the need to walk to get them even if one knows where they are are both forms of friction that can be addressed with the 5S workplace organization program.

Let's return to Taylor's (1911) pig iron handling example to illustrate the general difference between friction and opportunity costs. "We were surprised to find, after studying the matter, that a first-class pig-iron handler ought to handle between 47, and 48 long tons per day, instead of 12 and a half tons." He achieved this by having the workers pace themselves with prescribed intervals of work and rest, thus removing the friction and almost quadrupling the job" productivity. This change, like the one in the fabric folding operation, required no capital expenditures.

The principle of opportunity costs would, on the other hand, ask whether the job should require laborers to carry 92-pound (42-kilogram) pig iron at all or could instead use conveyors even though forklifts had yet to be invented. The roller conveyor was not invented until 1908, long after Taylor's "high-priced man" learned to carry 47 rather than 12.5 tons of pig iron a day, but farmers used conveyors to load grain onto ships as early as 1795 (Techspex, 2014). DoverMEI (no date given) elaborates that the 1795 conveyors used hand-cranked leather belts, but the Royal Navy was using steam-powered ones in 1804. *The Iron Age* (1897) has pictures of powered conveyors for pig iron at a steel mill, which raises the question as to whether similar equipment could have done the job of loading the iron onto rail cars. The gap between the much lower cost of using similar conveyors instead of human muscle to load pig iron onto trains is the *opportunity cost* of failure to adopt the superior alternative. Ford (Ford and Crowther, 1922) explained the basic concept as follows.

> If a device would save in time just 10 per cent. or increase results 10 per cent., then its absence is always a 10 per cent. tax. If the time of a person is worth fifty cents an hour, a 10 per cent. saving is worth five cents an hour. If the owner of a skyscraper could increase his income 10 per cent., he would willingly pay half the increase just to know how. The reason why he owns a skyscraper is that science has proved that certain materials, used in a given way, can save space and increase rental incomes. A building thirty stories high needs no more ground space than one five stories high. Getting along with the old-style architecture costs the five-story man the income of twenty-five floors. Save ten steps a day for each of twelve thousand employees and you will have saved fifty miles of wasted motion and misspent energy.

Taylor's "Improved" Pig Iron Handling Still Shows Enormous Waste

There is in fact a public domain (due to age) video of Taylor's improvement to pig iron handling that almost quadrupled productivity.[2] There is still enormous waste built into this job despite Taylor's improvement. The worker does lift with his legs rather than his back, but he is still lowering and raising his entire upper body weight along with the 92-pound pig iron. He also walks and carries the load, and Henry Ford would later point out that no job should require anybody to bend over or take more than one step in any direction.

The worker then carries the pig iron up a ramp. Imagine how much labor could be saved with an appropriate material conveyor here. The video also shows that the workers walk quickly up the ramp, and almost run, to reduce the amount of time they are under load. That is, the additional effort needed to go up the ramp quickly is more than offset by the reduced time under load.

The worker then sits down to rest as directed by Taylor and thus paces himself to avoid exhaustion. This is a good idea for everybody while shoveling snow, by the way. While nothing in this book constitutes medical advice, shoveling quickly to "get the job done" carries a real risk of a heart attack, especially if one is not used to this kind of vigorous work.

If we recall the advice to "think like a Greek," somebody like Archimedes or Alexander the Great would have probably questioned whether people should have carried the pig iron at all, even in the most work-efficient manner possible. This chapter mentioned previously that *The Iron Age* (1897) depicts powered conveyors for pig iron at a steel mill. National Iron and Steel Publishing Company (1905, 29) features an advertisement from Link-Belt Engineering for another pig iron conveyor whose operation is somewhat clearer (Figure 5.4).

This example illustrates both friction as removed from the original job by Taylor and the opportunity cost of not using a machine to take the backbreaking labor out of human hands entirely. None of this detracts from Taylor's very real achievement of enabling workers to do almost four times as much work for roughly the same amount of physical effort. The key takeaway is simply that neither Taylor nor anybody else seemed to question why this job used hand labor at all when steam, electricity, and internal combustion engine power were widely available. This principle carries over into modern agricultural jobs that are equally wasteful of labor. This chapter will show that this is often why food is expensive, farm workers get meager pay, and farmers little if any profit.

Suppose for another example that somebody like Taylor or Gilbreth had come up with a motion-efficient method to pick cotton by hand. Hand picking of cotton did require some skill,

Figure 5.4 Pig Iron Conveyor

and experts could work two or three times more rapidly than amateurs. Selection of the best motion-efficient method would have eliminated the friction associated with the job, but even a threefold productivity improvement in manual harvesting would not have been competitive with even John Rust's harvester from the 1940s let alone today's machines. The opportunity cost of not using a machine reflects the difference in question.

As yet another example, suppose somebody had found a more efficient way for the workers who dug the Suez Canal to use their hands for the excavation or made the very low-cost improvement of equipping them with shovels. This would have addressed the issue of friction in the form of motion inefficiency, but not the opportunity cost of failure to select steam power instead. William Otis's steam shovel (Figure 5.5) was patented in 1839, long before work on the Suez Canal began.

There are in fact pictures of steam-powered equipment, some on ships, working on the Suez Canal even though Emerson reported that hand labor was used. Recall that Ferdinand de Lesseps eventually brought in machinery to replace the unpaid labor of corvée, which explains the discrepancy between Emerson's observations and the apparently later presence of steam equipment.

Remember that poor quality is the only one of the Toyota Production System's Seven Wastes that announces its presence. The other six are (1) asymptomatic, (2) usually built into the job and thus present 100 percent of the time, and (3) usually much more costly than poor quality. This book will show that frontline workers are often in the best position to recognize friction that is built into their jobs, and gap analysis is another way to identify friction and/or opportunity costs.

Gap Analysis

A gap analysis is simply a comparison with the current performance state with an achievable performance state. Emerson (1909, 17) provides an outstanding example.

> Man wastes three-quarters of the coal in the ground, brings the remaining quarter to the surface by inefficient labor and appliances, doubles, trebles, or quadruples its

Figure 5.5 Otis Excavator, 1841

> cost by transportation charges to furnace door. Rarely is as much as 10 per cent of the energy of the coal transformed into electrical energy, and of this only 5 per cent can appear as light. Ten to twenty times as much light is provided as necessary on a writing table, because of the distance of the bulbs from the place where the light is needed. The light itself glows continuously, not only during intermittent work but often several hours before and after it is needed. Out of ten thousand B. t. u. in the coal mine we use in necessary light the equivalent of about six.
>
> The fire-fly converts the hydrocarbons of its food into light with an efficiency of 40 per cent. It flashes its light at intervals, thus making it most effective by contrast with the surrounding darkness, and it emits no more light than is necessary for its purpose.
>
> In production the fire-fly is about seven hundred and fifty times as efficient, in volume use ten times as economical, in time use twice as economical. The fire-fly is fifteen thousand times as efficient as his human rival.

Emerson realized more than a century ago that light production by bioluminescence is 40 percent efficient while his era's incandescent lights, as powered by electricity from coal-fired power plants, used only 0.06 percent of the energy in the coal. This did not tell him how to make lights more efficient but it did tell him it was possible because insects were doing it. The incandescent lights of 1909 converted only five percent of the electricity into light, and this improved to 20 percent subsequently (Indiana University of Pennsylvania, no date given). The same reference says that light-emitting diode (LED) bulbs, which convert electricity directly into light, are 90 percent efficient. There are also office and commercial lighting systems that detect occupancy and turn off the lights when nobody is present in the room. This addresses the issue cited by Emerson: "The light itself glows continuously, not only during intermittent work but often several hours before and after it is needed."

Emerson also identified the problem associated with all engines that seek to convert heat into energy. Recall that there is a hard upper limit on the amount of mechanical work that can be obtained by transferring heat from a hot reservoir such as a furnace to a cold one such as water in a condenser, and this is defined by the Carnot cycle. The Carnot cycle is theoretical, and those that apply to real-world applications such as the Rankine cycle for steam engines and the Otto cycle for combustion engines are even worse. Emerson (1909, 14) stated,

> An oil engine may reach 30 per cent thermal efficiency, but the salmon, assuming his whole weight to be pure oil, without consuming it, uses up several times more energy than is yielded by an equal weight of oil in combustion. The salmon uses atomic, not thermal, energy.

Emerson did not mean that salmon use nuclear power; "atomic" referred instead to chemical energy. The key observation is the fact that animals can convert chemical energy into mechanical (muscle) power very efficiently shows this to be possible and thus potentially achievable by human science. The fuel cell, which converts chemical energy directly into electrical power, is similarly not subject to the thermodynamic efficiency limits of the Carnot cycle. The US Department of Energy (no date given) states that while combustion power plants are still only 33–35 percent efficient, and gasoline engines are closer to 20 percent efficient, a fuel cell can achieve 60 percent efficiency. These examples show that identification of the gap between the current state and what is possible is the first step in the realization of the opportunity in question. *If somebody or something* (like a firefly or salmon) *is doing it, it can be done.*

More about Friction

While removal of friction can usually deliver improvements of "only" a few hundred percent—the Japanese gown production case study is an exception—in comparison to the selection of superior equipment, the latter must be available and large capital outlays might be required. As but one example, a major barrier to mechanized cotton harvesting was the fact that while barbed spindles could remove the cotton from the bolls, it was then difficult to get the cotton off the spindles. John Rust discovered that a moist spindle could strip the cotton fibers from the boll without this problem, and introduced a practical harvester in 1935 but did not have the resources with which to mass produce it. International Harvester commercialized a practical cotton harvester in 1944.

Another issue was the fact that the bolls matured at different rates, which meant that human judgment was still necessary to harvest them. Cotton growers therefore bred plants whose bolls matured at the same time to allow mechanical harvesting. This meant that the enormous productivity improvements that began with Rust's invention and evolved into the John Deere CP690 and its counterparts such as the Case IH Cotton Express were simply not available prior to roughly the 1940s. This meant that when cotton absolutely had to be harvested by hand, there was doubtlessly a "one best way" in terms of motion efficiency that would allow the worker to be paid more while the cotton could be sold for less. There were indeed differences between the productivity of skilled and unskilled cotton pickers, and the former doubtlessly commanded higher wages.

The same principle carries over into modern jobs, especially those not currently amenable to automation, where removal of all forms of friction can double or even quadruple productivity.

Fast-food jobs, home construction, roofing, landscaping, fruit and vegetable picking, and countless other occupations that involve manual labor, whether skilled or unskilled, are all amenable to this approach.

Modern Depictions of Friction

General von Clausewitz recognized the effect of friction on military operations, and Henry Ford (1930, 187) applied the same principle to industry.

> It is the little things that are hard to see—the awkward little methods of doing things that have grown up and which no one notices. And since manufacturing is solely a matter of detail, these little things develop, when added together, into very big things.

Tom Peters (1987, 323) applied the same concept to new product introduction. "The accumulation of little items, each too trivial to trouble the boss with, is a prime cause of miss-the-market delays."

Halpin (1966, 60–61) described friction in industry as follows. "They turned out to be the little things that get under a worker's skin but are never quite important enough to make him come to management for a change." A worker may have to walk to get tools and parts or bend over routinely as part of his or her job but, as the work gets done and the boss is happy, he or she never complains even though removal of the friction might easily double productivity.

Halpin is far from the only industry professional to apply the concept of Clausewitz's friction to workplaces. Shigeo Shingo (Robinson, 1990, 14) wrote, "Unfortunately, real waste lurks in forms that do not look like waste. Only through careful observation and goal orientation can waste be identified. We must always keep in mind that the greatest waste is the waste we don't see." Taiichi Ohno (1988, 59), the creator of the Toyota Production System, wrote similarly,

> In reality, however, such waste is usually hidden, making it difficult to eliminate ... To implement the Toyota Production System in your own business, there must be a total understanding of waste. Unless all sources of waste are detected and crushed, success will always be just a dream.

Emerson (1909, 59) added,

> It is not to be forgotten that in the human organism the whole is incapacitated by a seemingly slight injury to a single part. No man will work efficiently with a cinder in his eye or a splinter under his nail. Neither will a plant work efficiently if little things go wrong.

We now see a consistent pattern of "countless minor incidents," "accumulation of little items," "little things that get under a worker's skin," "little things that are hard to see," and "little things that go wrong" combining to undermine organizational performance enormously *especially because the friction is built into the job as it is currently performed.* Recognition of this principle empowers everybody in the organization to remove the friction to improve efficiency decisively and thus enable higher wages, lower prices, and higher profits simultaneously. Workers should understand clearly that if they experience even a seemingly minor inconvenience or problem that recurs over and over, they are not expected to "live with" or "work around" waste that is built into their jobs.

The Value-Adding "Bang!"

A good way to understand friction is in the context of what Masaaki Imai (1997, 22–23) calls the value-adding "Bang." "There is far too much muda [waste] between the value-adding moments. We should seek to realize a series of processes in which we can concentrate on each value-adding process—Bang! Bang! Bang!—and eliminate intervening downtime." There are some jobs in which the value-adding moment is a literal "bang" such as when a roofer's hammer or nail gun fastens a tile to a roof, or a punch press transforms a part. This value-adding moment is often a tiny fraction of the job's actual cycle time. When the roofer has to move material or, even worse, carry tiles up a ladder, he or she is not adding value and cannot be paid for the time thus spent.

The value-adding moment in bricklaying occurs when the mason sets a brick in a wall. If the mason has to bend over to pick up each brick, then he or she cannot really be paid for this effort either because the equivalent of 125 toe touches per hour does not add value for the customer. The value-adding "bang" is sometimes more like a whirring noise, such as that made by a drill, lathe, or other machine tool when it actually transforms the product. There are in fact five categories into which work activities can be classified:

1. Transformation of the product, which is the only value-adding activity.
2. Setup and handling such as putting a part into a fixture for machining, and then removing the finished work. It is vital to not allow setup and handling to masquerade as value-adding because they are not.
3. Inspection may be mandatory to ensure quality, but it does not add value.
4. Transportation, e.g., from one part of a factory to another, does not add value and has been addressed with moving assembly lines and work cells.
5. Delay contributes nothing, and "wait" is indeed a four-letter word even though we are allowed to say it on the radio.

It is probably a mistake to classify work as processing, inspection, transportation, and delay because the "processing" category gives setup and handling a perfect place to hide. Taylor (*Shop Management*, 1911) makes a clear distinction between transformation, and setup and handling as shown in Figure 5.6.

DeLuzio (2021) adds that in the context of standard work, setup time is not an allowable deduction from available time. The same goes for bathroom breaks noting that another operator should be available to take over to avoid a stoppage in production. This is particularly applicable for a capacity-constraining resource under Goldratt's Theory of Constraints because time lost at the constraint is lost forever. The bottom line is however that assessment of a process should put setup and handling, and product transformation, into separate categories so the former cannot masquerade as a value-adding task.

Friction, Motion Efficiency, and Standard Work

This book's evident fascination with soldiers and weapons stems from the fact that being shot at is a powerful incentive to figure out how to shoot back as rapidly as possible. A civilian business can survive for decades or even centuries with inefficiencies of 90 percent or greater, at least until a competitor makes a breakthrough. The trade of bricklaying apparently tolerated, as proven by Frank Gilbreth, a 64 percent inefficiency for centuries and nobody thought to do anything about it. In the movie version of Bernard Cornwell's *Sharpe's Eagle*, starring Sean Bean as Richard

SHOP MANAGEMENT 171

THE MIDVALE STEEL CO.

Form D—124. Machine Shop.......................18.........

ESTIMATES FOR WORK ON LATHES

OPERATIONS CONNECTED WITH PREPARING TO MACHINE WORK ON LATHES AND WITH REMOVING WORK TO FLOOR AFTER IT HAS BEEN MACHINED

OPERATIONS	TIME IN MINUTES	
Putting chain on, Work on Floor		
Putting chain on, Work on Centers		
Taking off chain, Work on Floor		
Taking off chain, Work on Centers		
Putting on Carrier		
Taking off "		
Lifting Work to Shears		

NAME ....
Sketch .... Number............
Order.............. Weight.............
Metal.............. Heat No. ....
Tensile Strength.... Chem. Comp.......
Per cent. of Stretch ....
HARDNESS, Class ....

OPERATIONS CONNECTED WITH MACHINING WORK ON LATHES

OPERATIONS	Speed	Feed	Cut	Tool	Inches	Minutes
Turning Feed In						
" " "						
" Hand Feed						
" " "						

Figure 5.6 Transformation versus Setup and Handling. Estimates for Work on Lathes. The left column depicts setup and handling, and the right column machining.

Sharpe, Sharpe explains to a badly led and badly trained regiment of inexperienced men that, as matters stand, they can fire two rounds a minute while the French can fire three. That means that unless they learn the rapid-firing drill that Sharpe can teach them, they will soon all be dead. This provided an immediate and compelling incentive to participate enthusiastically in Sharpe's counterpart of a kaizen event.

Military organizations were therefore well ahead of civilian ones in recognition of the fact that motions that do not add value are waste and that neither soldiers nor their weapons could afford non-value-adding tasks. *The following story is absolutely true except for the recruiting sergeant's knowledge of the success secret in question* (Figure 5.7). Had he known it, he might have easily become the wealthiest man in England.

Recruiting sergeants were known for a wide variety of imaginative sales pitches,[3] and this one proclaimed, "What would you say, lad, if I told you that service in King George's army will put the secret of almost limitless wealth into your hands?"

"I wouldn't call sixpence a day with stoppages for 'necessaries' limitless wealth," the young man replied, "but times are hard and I can't find civilian work, so I'll serve the King for three meals a day." The Army's role as the only alternative to hunger and unemployment is reflected meanwhile in the folk song "Marching through Rochester," "The Rochester Recruiting Sergeant," "The Bold Fusilier," or the "Gay Fusilier" (all are similar and are set to the same music as "Waltzing Matilda"). None of the men with trades such as butcher or baker will take Queen Anne's shilling to fight for the Duke of Marlborough, and the only man who does has often stood in "the parish queue," i.e., the bread line. Figure 5.8 shows, in fact, that recruiting sergeants of the 33rd Regiment of Foot used oatcakes, also known as havercakes, to entice hungry men into the ranks.

Figure 5.7 The Recruiting Sergeant (by John Collet, 1757)

Figure 5.8 The Havercakes' Recruiting Sergeant. Walker, George. 1814. "A Recruiting Sergeant of the 33rd Regiment of Foot, 1814." It's in the public domain due to age, and can be found from multiple sources. The National Army Museum (UK) may be cited as a source. https://collection.nam.ac.uk/detail.php?acc=1961-10-67-1

Once he enlisted, the recruit would have then had to master a loading drill similar to that written by von Steuben (1779) for the United States Army. Steps I through IV are related to aiming and firing the musket, so the soldier begins with an empty musket for Step V. The resemblance to a modern *job breakdown sheet* is evident. There is a prescribed series of actions, each of which incorporates the best-known motion-efficient way to do it, along with the number of motions required for the purpose.

> V. *Half cock—Firelock!* One motion.
>
> Half bend the cock briskly, bringing down the elbow to the butt of the firelock.
>
> VI. *Handle—Cartridge!* One motion.
>
> Bring your right hand short round to your pouch, slapping it hard, seize the cartridge, and bring it with a quick motion to your mouth, bite the top off down to the powder, covering it instantly with your thumb, and bring the hand as low as the chin, with the elbow down.
>
> VII. *Prime!* One motion.
>
> Shake the powder into the pan, and covering the cartridge again, place the three last fingers behind the hammer, with the elbow up.
>
> VIII. *Shut Pan!* Two motions.
>
> Shut your pan briskly, bringing down the elbow to the butt of the firelock, holding the cartridge fast in your hand. Turn the piece nimbly round before you to the loading position, with the lock to the front, and the muzzle at the height of the chin, bringing the right hand up under the muzzle; both feet being kept fast in this motion.
>
> IX. *Charge with Cartridge!* Two motions.
>
> Turn up your hand and put the cartridge into the muzzle, shaking the powder into the barrel. Turning the stock a little towards you, place your right hand closed, with a quick and strong motion, upon the butt of the rammer, the thumb upwards, and the elbow down.
>
> X. *Draw Rammer!* Two motions.
>
> Draw your rammer with a quick motion half out, seizing it instantly at the muzzle back handed. Draw it quite out, turn it, and enter it into the muzzle.
>
> XI. *Ram down—Cartridge!* One motion.
>
> Ram the cartridge well down the barrel, and instantly recovering and seizing the rammer back handed by the middle, draw it quite out, turn it, and enter it as far as the lower pipe placing at the same time the edge of the hand on the butt end of the rammer, with the fingers extended.
>
> XII. *Return Rammer!* One motion.
>
> Thrust the rammer home, and instantly bring up the piece with the left hand to the shoulder, seizing it at the same time with the right hand under the cock, keeping the left hand at the swell, and turning the body square to the front.

Nobody wanted to take the time to measure out the powder from a powder horn in the middle of a battle because this was a form of handling that took time away from the all-important and literal value-adding "Bang!" The paper-wrapped cartridge, with its premeasured powder charge along with the bullet and the wrapper that served as the wadding, exemplified what we now call *external setup*. When soldiers used matchlock muskets and arquebuses (Figure 5.9), they carried premeasured charges in wooden containers for the same reason.

It is also to be noted that the soldier was told to bite the paper-wrapped cartridge open because this took far less time than, for example, resting the musket on the ground to use both hands for this purpose. Some Russian peasants knocked out their own front teeth because they regarded conscription into the Tsar's or Tsarina's army as the equivalent of a death sentence, noting especially that most deaths were from disease or privation rather than combat, and Russia's counterparts of press gangs would not take men who could not bite open a cartridge.[4] This aspect of the loading drill also incited the Sepoy Mutiny when a rumor spread that the British were greasing their cartridges with pig and cow fat, which were forbidden to Muslims

Figure 5.9 Matchlock Arquebusier. Jacob de Gheyn II, 1607. "Instructions for the use of the musket," public domain due to age. The cylinders that hang on the soldier's bandolier are wooden cartridges, each of which holds a premeasured charge for the musket. This is a form of *external setup* in which the job of measuring the correct charge is removed from the immediate operation

and Hindus respectively. The question arises as to whether a fixture could have been installed next to the musket's priming pan to tear off the end of the cartridge and right next to where it was first needed, which could have eliminated the motions necessary to bring the cartridge to the mouth and then down to the pan along with any concerns over the lubricant on the cartridge.

Recall that Bernard Cornwell's protagonist Richard Sharpe becomes a motion efficiency expert in *Sharpe's Eagle.* He tells soldiers how they can omit the operations of drawing the ramrod, ramming down the cartridge, and returning the ramrod by slamming the butts of their muskets against the ground to drive the charge and the bullet to the back of the barrel. This might or might not be realistic, although various reenactment sources do refer to tap loading. The *thought process* is however the key takeaway. "Draw rammer" (two motions), "Ram down cartridge" (one motion), and "Return rammer" (one motion), for a total of four motions, are replaced by the single motion of slamming the musket's butt against the ground. The effect is to reduce the setup time between the value-adding and literal "Bangs."

The question also arose as to whether the motions necessary to prime the pan could be eliminated, and there were in fact self-priming muskets during the eighteenth century. The Austrian M1784 musket was apparently self-priming, and there is a video (SBAM, 2016) of a reenactor who bites open the cartridge, omits the priming step, pours the powder down the barrel, and rams the charge home to fire six rounds in 70 seconds. The Springfield M1855 Rifle-Musket meanwhile used the Maynard tape primer system (Figure 5.10), which still appears in children's cap guns. This eliminated the need for the soldier to use an additional motion to apply a new percussion cap for each round. This underscores the kind of thinking that led to more efficient weapons during the horse and musket era and carries over into modern industries. *If the process step doesn't add value, figure out how to get rid of it.*

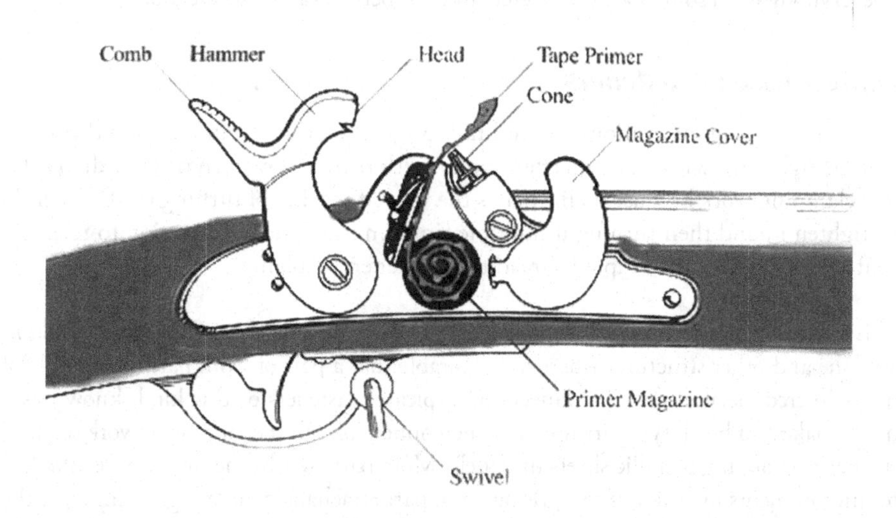

Figure 5.10 Maynard Tape Primer. US National Park Service, no date given. (Public domain as a publication of the US Government)

Our recruit is now a trained soldier, and he returns home after many years of service to enter a civilian trade. Where, he wonders, is this secret of limitless wealth the recruiting sergeant promised him long ago? It was not until the twentieth century that Frank Gilbreth (1911), the father of motion efficiency, provided the answer.

> The United States government has already spent millions and used many of the best of minds on the subject of motion study as applied to war; the motions of the sword, gun, and bayonet drill are wonderfully perfect from the standpoint of the requirements of their use. This same study should be applied to the arts of peace.

This section promised that the story was true except for the recruiting sergeant's knowledge of the secret in question, which was actually delivered by Gilbreth.

Another important takeaway from this story is that tens or perhaps even hundreds of millions of soldiers learned these drills during the horse and musket era and even the horse and rifle era of the late nineteenth century. It was only in the early twentieth century that anybody (Gilbreth, and also Emerson as he depicted loading drills on American battleships) realized that the same principles were applicable to civilian occupations. Napoleon's observation that every French soldier carried a marshal's baton in his backpack, for promotion was available to anybody who merited it, is equally applicable here because *any of the countless soldiers who returned to civilian occupations after learning these principles had, if he had only realized it, the power to transform the entire world* (Figure 5.11). Any hourly worker who similarly comprehends these principles has today the power to realize enormous productivity gains in his workplace, so the lesson is to never assume that your job's design is the best possible.

Standard work can also address variation in the time necessary to complete each task—the same variation that resulted in large quantities of inventory in Goldratt's and Cox's (1992) *The Goal*. Stern (1939, 5–6) pointed out some enormous variation in worker productivity in shoe manufacture. One worker needed 82.5 seconds to stitch a circular seam, while another needed only 46. This proved that the job could be done in 46 seconds, and the next question would be as to how the worker managed to do it so quickly. Application of Taylor's "one best way" principle would have made this the standard for everybody until somebody discovered an even better way to do the job.

Interrupted Thread Fasteners

Many other time-saving innovations had military origins, such as the interrupted thread fastener, which can be tightened with a one-quarter or one-sixth turn of a wrench or screwdriver. It is very useful for clamping work because it eliminates the wasted motion of turning the fastener multiple times to tighten it, and then turning it multiple times in the opposite direction to release it later. US Patent US2850782A, Interrupted thread type fasteners, explains,

> This invention has to do with improvements in fastening devices of the type usable in airplane and other structures where it is desirable that a pair of work parts be capable of quick interconnection and disconnection. Typically, fasteners of this kind, known as a quick make and break type, are used for interconnection of inner and outer work sections such as overlapping metallic sheets or panels. More particularly, the invention is directed to improvements in fasteners comprising a nut part attachable to one, e.g. an inner, of the work parts or sheets, and a stud part engageable against the other or outer work part or sheet, the stud and nut having interrupted threads so that the stud is receivable by axial movement into the nut and is threadedly engageable therewith upon rotation.

Figure 5.11 A Motion Efficiency Expert (Had He Only Realized It). Soldier of the 29th Regiment of Foot, 1742. (Public domain due to age)

The De Bange breech obturator system (Figure 5.12), which was invented in 1872, used similarly an interrupted thread breech block to close a breech-loading artillery piece with a partial turn of the operating handle noting (again) that having to turn the block multiple times to effect a seal would be far too time-consuming.

The key takeaway here is that a worker adds value to a product only when he or she transforms it through machining, assembly, or a similar action. Even turning a screw or a bolt adds value only with the final turn that tightens it, as most of us know from home improvement and/or do-it-yourself furniture assembly projects. Screws seem to go round and round almost forever, and especially if one is tightening them by hand, which convinced me to buy a powered screwdriver for household jobs that involve less than a dozen screws as opposed to a full-time assembly job. Setup,

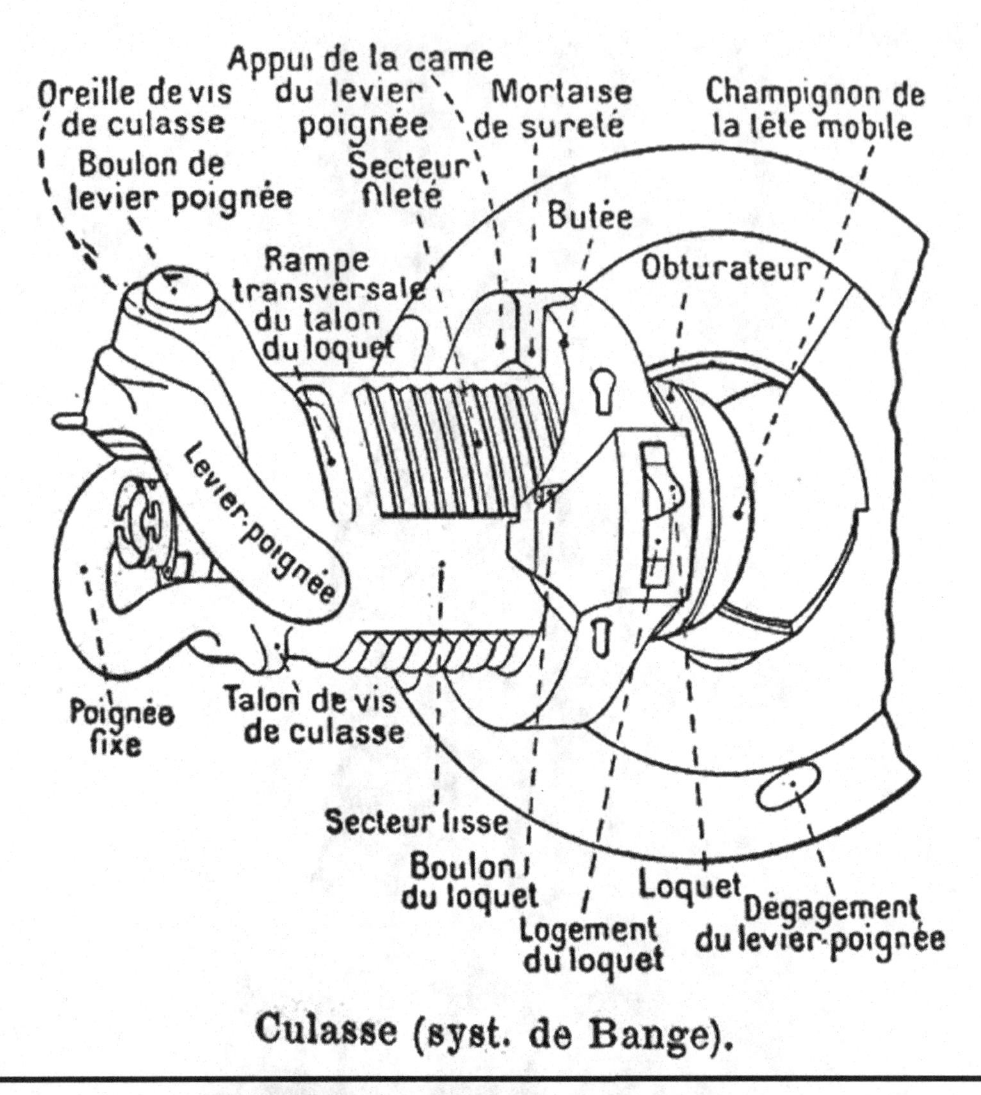

Figure 5.12 De Bange Breech Obturator. De Bange breech obturator system. Modern interrupted thread fasteners and clamps use the same principle (*Encyclopedie Larousse Illustree*, 1897. Public domain due to age.)

handling, transportation of materials, parts, and tools, and of course the need to wait for anything do not add value, and the same goes for unnecessary motions; remember that only the final turn of the fastener adds value. Removal of this kind of waste from a job enables higher wages, lower prices, and higher profits simultaneously.

Japanese Disposable Gowns

Toyota Times Global (2020) reports how a factory had to increase its output of disposable gowns to address the COVID-19 epidemic. Application of the Toyota Production System's methods, along with items purchased from a 100-yen (roughly one dollar) store, *increased productivity 100-fold.* An inexpensive cardboard tube was used to smooth the material, as opposed to doing it by hand. Scissors were kept in a single location (which is consistent with the 5S workplace organization program), and a light box was used to help with defect inspection. Much of the work is, in contrast to automobile manufacture and mechanized cotton harvesting, labor-intensive, but the labor was nonetheless made far more efficient. This example is also a rare exception to the general rule that 100-fold improvements usually tend to require at least some automation.

Fruit Harvesting

Workers who pick fruit waste an enormous amount of time and labor by climbing ladders to do their jobs. Somebody cut the end of a plastic bottle to make a sharp edge and attached the bottle to a pole with which to reach up into a tree to harvest the fruit without climbing. A modification added a soft cloth chute to the bottle to deliver the fruit to the user without the need to lower the pole and reach into the bottle each time. Similar devices are now available commercially and can be purchased for as little as $20. This is what one might use to double or triple productivity if one cannot afford, for example, drones that use machine vision to identify fruit that is ready to harvest and a mechanical arm to collect it. The instant anybody sees a worker climb a ladder in an orchard, however, this observation should be cited immediately as friction.

Shoveling

Taylor (1911) described how friction was built into the simple labor-intensive job of using a shovel to move materials. The ideal shovel load is roughly 21 pounds, but the actual load is the capacity of the shovel multiplied by the density of the material in question, and materials are of course not equal.

> At the works of the Bethlehem Steel Company, for example, as a result of this law, instead of allowing each shoveler to select and own his own shovel, it became necessary to provide some 8 to 10 different kinds of shovels, etc., each one appropriate to handling a given type of material not only so as to enable the men to handle an average load of 21 pounds, but also to adapt the shovel to several other requirements which become perfectly evident when this work is studied as a science. A large shovel tool room was built, in which were stored not only shovels but carefully designed and standardized labor implements of all kinds, such as picks, crowbars, etc. This made it possible to issue to each workman a shovel which would hold a load of 21 pounds of whatever class of material they were to handle: a small shovel for ore, say, or a large one for ashes. Iron ore is one of the heavy materials which are handled in a works of this kind, and rice coal, owing to the fact that it is so slippery on the shovel, is one of the lightest materials. And it was found on studying the rule-of-thumb plan at the Bethlehem Steel Company, where each shoveler owned his own shovel, that he would frequently go from shoveling ore, with a load of about 30 pounds per shovel, to handling rice coal, with a load on the same shovel of less than 4 pounds. In the one case, he was so overloaded that it was impossible for him to do a full day's work, and in the other case he was so ridiculously underloaded that it was manifestly impossible to even approximate a day's work.

We take the same issue for granted today when we select a large snow shovel—and only recently have ergonomic designs become available that enable far better results for the same amount of effort—to move powdery snow and a smaller one to move wet snow. The fact that people have shoveled snow for decades without the benefit of ergonomic shovels shows, however, widespread inattention to the issue of friction.

Bricklaying and Roofing

We have already seen how Frank Gilbreth's introduction of a non-stooping scaffold, which delivered bricks at waist level, increased productivity from 125 to 350 bricks per hour. Taylor (1911) described the friction that was built into the original job as follows.

> Think of the waste of effort that has gone on through all these years, with each bricklayer lowering his body, weighing, say, 150 pounds, down two feet and raising it up again every time a brick (weighing about 5 pounds) is laid in the wall! And this each bricklayer did about one thousand times a day.

Bricklaying is among the world's oldest trades, *which means this waste of labor hid in plain sight for thousands of years.* This underscores yet again the proposition that poor quality is the only Toyota Production System waste that announces its presence, while the others are (1) asymptomatic, (2) usually built into the job, and (3) usually more costly than poor quality. Even though Gilbreth and Taylor identified this kind of waste more than a century ago, the lessons learned have clearly not been deployed into many of today's construction and home improvement jobs. There are photos and videos of roofers carrying heavy bundles of tiles up ladders, which requires a lot of physical effort and could conceivably—nothing in this book constitutes formal engineering or occupational safety advice—create a fall hazard in the bargain.[5] More to the immediate point is that instead of having to bend over for each brick, the roofer must climb down and up a ladder for each bundle of tiles which is just as wasteful if not even worse. Mace Industries' Bumpa Portable Conveyor Tile Hoist and E-ZLift's Chain & Flight Towable Elevating Carriage (as but two examples) deliver the tiles so the roofers can focus on the value-adding task of building a roof as opposed to carrying heavy loads. The concept is hardly new, however, because people in medieval times used treadmill cranes (Figure 5.13) to move heavy loads from ground level to the construction workers.

If one takes a close look at this illustration of a treadmill crane, a couple of other items should be clear to anybody who knows what to look for. *The bricks appear to be delivered to the masons at waist level, as recommended by Gilbreth more than 700 years later.* One might ask why people who were smart enough to not only use a treadmill crane to deliver bricks to the point of construction but also deliver them at waist level had a man on a ladder with a container of mortar on his back. We can speculate intelligently that his function was to move up the ladder as the height of the wall increased to keep the mortar at, or at least near, waist level for the mason as well. This is yet another example of a work efficiency "secret" that hid in plain view for literally centuries because as Sir Arthur Conan Doyle's Sherlock Holmes put it, "You see, but you do not observe." Remember that watching the value-adding workplace (gemba) from Taiichi Ohno's circle is useful only if one knows what to look for.

Floor Tiles, Sidewalks, and Safety Tape Marking

Workers normally lay tiles one at a time, but there was a video on LinkedIn of a worker using a frame that accommodated six tiles at a time and ensured their correct spacing. The job went

Figure 5.13 Treadmill Crane, Thirteenth Century. Construction of the Tower of Babel in the Maciejowski Bible (public domain due to age)

very quickly. Another video shows workers placing a template or mold on top of wet concrete or a similar material, and stepping on it to form a pattern similar to that of hand-laid bricks. They achieve the desired result perhaps ten times as rapidly as they would by kneeling to place actual bricks.

There are also pictures and videos of workers kneeling and having to move around in this position, to attach safety or other marking tape to floors. Another video showed a pole similar to that of a mop or a broom, but with two spools of tape at the bottom, that enabled the worker to lay two parallel rows of tape simultaneously while walking. This is how relatively cheap and simple tools can deliver efficiency improvements of several hundred percent which means the job costs the employer less while the workers can be paid a lot more.

Standard Work

Recall that standardization does not mean that workers are to leave their brains at the factory gate. A standard such as a work instruction contains the best-known way to do a job, which can be superseded by a better way as soon as it is discovered and proven to work. It ensures consistency in the job's output and also prevents backsliding to previous, and inferior, methods.

Standards Are Documented

Standard Work for the Shopfloor (2002, Productivity Press Development Team) provides an excellent overview of standard work, which can play a central role in eliminating friction from the workplace. First (p. 2), standards are documented. Written records prevent loss of knowledge and also avoid reliance on word of mouth or individual opinions as to how a job is to be performed. It is to be remembered that the Roman push scythe was largely forgotten for centuries even though its description was written down but apparently not where the information was readily accessible.

More to the point, the Ford Motor Company's phenomenal efficiency methods were largely forgotten for decades after the deaths of Henry and Edsel Ford and the retirement of Ford's production chief Charles Sorensen. Ford's books (*My Life and Work, Today and Tomorrow,* and *Moving Forward*) described exactly what he did but these were apparently not part of the company's documented information system. When the company rediscovered them, it believed them to be Japanese in origin. Some Ford executives visited Japan in 1982 to learn about the purportedly Japanese efficiency methods. "One Japanese executive referred repeatedly to 'the book.' When Ford executives asked about the book, he responded: 'It's Henry Ford's book of course—your company's book'" (Stuelpnagel 1993, 91). Organizations should therefore adopt the position that if it isn't written down, it didn't happen or might as well not have happened.

Standard Work for the Shopfloor (2002, 20) adds that standards should ideally be only one page in length, and they should be easy to understand. The example given on page 21 includes the operation's cycle time, the time for each operator, and also the percent load for each operator. Graphical descriptions are provided as to what is to be done and which operator is to do it. Another figure shows the location of the incoming and exiting work and the movement of the operators.

Workers are instructed to not tolerate friction, such as the need to fix or correct the process repeatedly for minor equipment failures. There is no such thing as an equipment failure or stoppage that is too minor to invoke corrective and preventive action (CAPA) to remove its root cause. Containment of the problem, in the form of repeated "fixing," is not corrective action. General Curtis Lemay said to stop swatting flies (containment of the symptoms, defects, or failure effects) and get rid of the manure pile (the root cause) that was attracting them.

Use the Standard to Identify Improvement Opportunities

The reference (p. 28) includes a standard operation sheet for a soldering operation whose first two steps are to cut the lead wires and remove their sheaths. This requires the worker to handle two tools, a wire cutter and a wire stripper, and they require 35 seconds of the 330-second total. My own inclination would be to ask whether wires could be supplied pre-cut and pre-stripped. The next step is to apply the solder flux and solder to the wires, which invites the question as to whether this step also can be externalized to save another 30 seconds. An Internet search shows that pre-cut and pre-soldered jumper wires are in fact commercially available. The reference (p. 64) discusses other improvements such as stabilizing jigs to eliminate unnecessary work.

Always classify each step as (1) value-adding transformation, (2) setup and handling, (3) transportation, (4) inspection and testing, or (5) waiting. Only the first adds value. Cutting and stripping the wires and application of flux and solder are technically transformation but could also be construed as a form of setup, i.e., preparing the wires for assembly, that might be subject to externalization, i.e., performance offline and outside the work cell.

Standard Work for the Shopfloor (p. 52) features two additional examples of reduction of waste motion by positioning inputs such as plastic bags for packaging and parts so they are ready at hand, as opposed to requiring the worker to fetch them. This is consistent with Henry Ford's (Ford and Crowther, 1922) guidance,

> Use work slides or some other form of carrier so that when a workman completes his operation, he drops the part always in the same place—which place must always be the most convenient place to his hand—and if possible have gravity carry the part to the next workman for his operation,

and it applies to incoming as well as outgoing work.

Elements of Standard Work

Standard work consists of:

1. *Takt time*, or the reciprocal of the production rate required to meet customer demand. The latter might be less than the operation's capacity because we generally don't want to produce items for which there is no immediate need. If it is more than the operation's capacity, then the operation is a constraint, bottleneck, or capacity-constraining resource (CCR).
   - Suppose for example that the operation can make 60 parts per hour, and the downstream customer needs 30 per hour. The takt time is therefore two minutes because one part every two minutes will meet the demand, even though the operation can make one part per minute. If the downstream operation needs 120 parts per hour, the takt time is 30 seconds, but this exceeds the operation's capacity, so the operation becomes the process constraint.
   - Recall that von Steuben's musket drill included takt time as defined by the number of motions required to perform each step. Takt is related to the German word for an orchestra conductor's baton, i.e., something that sets a time or tempo for an activity, and armies used drums to coordinate activities that ranged from loading and firing muskets to marching in step. A Greek trireme's keleustes or a Roman warship's hortator beats a drum to synchronize rowing by citizens in the former case and slaves in the latter, as depicted in *Ben Hur* with Charlton Heston. Sailors used songs known as shanties not just for entertainment while they worked, but to set a tempo for activities such as pulling ropes to set sails.
   - *Cycle time* is the total time between the time work enters a process and the time it leaves. *Standard Work for the Shopfloor* (2002, 37) adds that if this can be reduced to takt time, single-unit flow is possible. DeLuzio (2021) adds that cycle time equals operator cycle time plus machine cycle time, i.e., the time necessary for the operator to load and unload the machine plus the time necessary for the machine to perform its task.
   - Loading and unloading the machine are non-value-adding handling whose reduction will not only increase capacity but also free operators to perform value-adding work. In addition, per Little's Law, inventory equals throughput (e.g., parts per minute) multiplied by cycle time (e.g., minutes) so less cycle time reduces inventory, one of the Toyota Production System's Seven Wastes, in the bargain.
2. *Standard work sequence* (p. 39 of the reference) is the order in which the operations in a process are to be performed. This is entirely consistent with the process orientation of modern operations management and also the process flowchart. The process sequence is not, however, necessarily the same as the work sequence because the latter may depend on the

number of workers in a work cell at a given time. The work sequence defines what each operator does, as opposed to what the process does. The reference adds, "Standard work sequences should be created for every possible combination of workers in a given cell."

  - DeLuzio (2021) adds that rework should not be performed in the work cell because rework is an abnormal situation that should be handled offline to avoid impacting takt time. Another way of saying this is that rework should be externalized, just as certain forms of setup should be externalized if possible. A material handler or "water spider" moves parts and materials between work cells for the same reason; operators should not need to leave their workstations to do this.

3. *Standard work-in-process (WIP) inventory* is the amount of inventory necessary to ensure continuity of operations. The concept is similar to that of the buffer in Goldratt's drum-buffer-rope (DBR) production control system in which a buffer of WIP is kept available at or moving toward the capacity-constraining resource to ensure that it never runs out of work. Buffers are not kept for operations with excess capacity in the DBR system because, unlike the constraint, they can make up any lost time. It might however be necessary to maintain standard WIP for processes in a kanban system where capacity is relatively balanced and any operation can become a constraint.
   - De Luzio (2021) defines the *curtain quantity* as the lead time for process inputs divided by the process takt time. If for example the lead time is 60 minutes, and the takt time is two minutes, then the curtain quantity is 30 units. Another way of saying this is that if 30 units are delivered every hour, they will meet the requirements of the takt time.

Another deliverable of standard work (*Standard Work for the Shopfloor*, 2002, 41) is a balanced workload in which no operator has too much or too little work to do. If operators have too little work to do, the work cell will probably have more people than it needs. If an imbalance leaves others with too much to do, they will not be able to meet the takt time. Ford (Ford and Crowther, 1922) cited the ideal situation roughly 100 years ago: "The idea is that a man must not be hurried in his work—he must have every second necessary but not a single unnecessary second." This does not mean the job's tempo should require workers to struggle to keep up but, on the other hand, people usually prefer to do something rather than wait for the next part to come along. There is not enough time between parts to allow people to even begin to do anything else, which means it is wasted time for the worker as well as the employer.

Standard work therefore suppresses friction by specifying (1) what is to be done, (2) the one best-known way to do it, and (3) exactly how quickly it is to be done. Takt time means that no operation gets ahead or falls behind by producing too much or too little, respectively, or by producing at the wrong time. Everybody works at the same pace (as set by the takt time) and does the job the same way, which eliminates variation in both quality and rate of production. There is not enough room in this book to go into all the details, but this section has hopefully provided a good overview, and readers who want to know more should refer to *Standard Work for the Shopfloor*.

The Job Breakdown Sheet

Standard work is synergistic with the job breakdown sheet, which is a three-column tabulation of (1) the process step, or what to do, (2) the important points, or how to do it, and (3) the reason, or why it is done.

The job breakdown sheet can be expanded to include a *job safety analysis* to support a workplace safety program and the ISO 45001 standard for occupational health and safety management systems. The job safety analysis is typically a three-column form that depicts (1) the job step, (2) the hazards, and (3) the controls whose purpose is to mitigate or eliminate the hazards. The second two columns (hazards and controls) can be easily appended to the job breakdown sheet. Quality considerations can be added as well to make the entire document a powerful tool for not only directing the job but also improving safety and quality. This can be extended even further to process failure mode effects analysis (PFMEA) which addresses not only quality but also safety and continuity of operations.

Standard work and job breakdown sheets are useful for identifying improvement opportunities on the order of up to a few hundred percent, but real breakthroughs come from attention to opportunity costs. These relate to gains not realized due to failure to implement new technologies and new forms of machinery.

Opportunity Costs

Taylor's pig iron example showed that there is an upper limit on the potential improvement from the elimination of friction alone. Motion efficiency can deliver threefold improvements (e.g., in bricklaying, as proven by Frank Gilbreth) or even fourfold (in pig iron handling), but there is an upper limit on how much work a person can actually achieve without the best available tools and working conditions. Attention to opportunity costs delivers tenfold and hundredfold improvements.

Textiles and Cotton

Ford described how the absence of an improvement is an invisible tax, for the cost accounting system cannot even acknowledge it, on the performance of the business. A gap between the current performance state and a potential one is a tax on the performance of the business. As but one example, the use of hand labor to harvest cotton when simple and inexpensive handheld machines will do the work of several hand laborers, and automated harvesters far more, can constitute an inefficiency ranging from 75 to 99.9 percent.

Crow (1943, 37) reported similarly how automation increased the productivity of textile workers 200-fold almost a century before Ford introduced the moving assembly line. The price of the cloth therefore consisted almost exclusively of the cost of the raw materials and the textile machines, and *cheap labor cannot offset material and capital costs.*

Anybody who believes, meanwhile, that an attempt to use cheap labor as a substitute for capital will have a happy ending is doomed to extreme disappointment. John Rust's original cotton-harvesting machine, which was invented roughly 80 years ago, could pick only one row of cotton which made it equal to forty hand laborers. Newer versions can harvest and process six or more rows of cotton at a time. John Deere's CP690 Cotton Picker can harvest ten acres of cotton per hour (Steadman, 2014).

Ganzel (2007) reports that it required 125 hours of manual labor to harvest one acre of cotton, which, in comparison to the CP690's performance, makes the latter 1250 times as productive. The CP690 also does more than just pick the cotton; it bundles it into modules. Had such machinery, or even the original International Harvester machines, been around in the mid-1800s, they would have put Simon Legree, the villain of Harriet Beecher Stowe's *Uncle Tom's Cabin*, out of business in very short order. Whirlston (no date given), a cotton-processing equipment manufacturer, contends, "Cotton is an extremely labor intensive crop, and this can be one reason that slavery existed" and adds that a cotton-harvesting machine was invented in 1850. It could not apparently distinguish, as John Rust's machine could do roughly 90 years later, between mature cotton bolls ripe for harvest and immature ones that needed to be left alone. Had it been able to do so, it might have in fact abolished slavery through pure economic force. Carlson (1974) asks, "If the South had had a cotton harvester would slavery gradually have disappeared; could the Civil War have been avoided?"

It is difficult to see how robot or corvée survives in Uzbekistan (Radio Free Europe/Radio Liberty, 2013) when advanced nations have machines of this nature. If people cannot afford the John Deere CP690's mid-six-figure price tag, India has handheld machines that resemble handheld vacuum cleaners that make one worker equal to three or four hand laborers. Rajkumar Agro Engineers Pvt Ltd sells a 12-volt device that weighs 700 grams or roughly 1.5 pounds that increases productivity from 50 to 200 percent, although the worker must still walk. Ganapathy Agro Industries sells a similar device on Indiamart—it looks like something one could easily buy in a home improvement store, except it picks cotton—for 5500 rupees which equals roughly US$75. It can harvest 125 kilograms of cotton per day versus the 20-kilogram robot or corvée quota in Uzbekistan. Another way of saying this is that even a $75 handheld machine can make one Indian worker equal to six Uzbeks who lack the machine, while the CP690 makes one American worker equal to more than a thousand Uzbeks. If we remember that unpaid workers must still receive food and other subsistence, a paid Indian worker is still cheaper than six unpaid Uzbeks and a high-wage American worker plus the cost of ownership of a $600,000 machine is far cheaper than a thousand unpaid Uzbeks. This underscores Aristotle's observation that automation would render slavery and variants like robot and corvée and by implication low-wage labor, obsolete.

Indiamart also features the Labh Group Battery Cotton Picker Harvester for six lakhs (hundred thousands) rupees or roughly US$8,155. This one, while far smaller than the CP690, allows the operator to ride in a cab, has a 120-liter fuel tank and can hold up to 19 cubic meters of cotton.

The key takeaway is again that when a machine can do the work of more than 1000 hand laborers, the employer can pay the machine's operator very high wages without a significant effect on the cost of the product. When the per-unit labor becomes negligible, as it does when the harvester's operator can pick an acre of cotton in six minutes, the contest ceases to be between high-wage labor and cheap labor, but machine against machine. The worker with the handheld machine outclasses several workers who must use their fingers, one worker with the simple riding machine outclasses the workers with the handheld machines, and so on.

Opportunity Costs in Agriculture

We have already seen how Uzbekistan uses unpaid hand labor to pick cotton although workers with even the cheapest machines would be more economical. There are widespread complaints about the low and even sub-minimum wages that are paid to the farm workers who harvest crops in the United States. Student Action with Farmworkers (2012) claims that, as of that date, the average annual income of farm workers was only $11,000 and also that these workers are not covered by minimum wage laws. Consumers, however, face routinely high prices for fresh blueberries, strawberries, and other fruits in grocery stores, while farmers struggle to stay in business. This reinforces the contention that *low wages are generally symptomatic of high prices and low profits*, instead of the low prices and high profits we might expect. Henry Ford (Ford and Crowther, 1922), who grew up on a farm, identified the reason roughly 100 years ago (emphasis is mine),

> The farmer makes too complex an affair out of his daily work. I believe that the average farmer puts to a really useful purpose only about 5 per cent of the energy that he spends. If any one ever equipped a factory in the style, say, the average farm is fitted out, the place would be cluttered with men. The worst factory in Europe is hardly as bad as the average farm barn. Power is utilized to the least possible degree. Not only is everything done by hand, but seldom is a thought given to logical arrangement. A farmer doing his chores will walk up and down a rickety ladder a dozen times. He will carry water for years instead of putting in a few lengths of pipe. His whole idea, when there is extra work to do, is to hire extra men. He thinks of putting money into improvements as an expense. Farm products at their lowest prices are dearer than they ought to be. *Farm profits at their highest are lower than they ought to be. It is waste motion—waste effort—that makes farm prices high and profits low.*

While mechanized agriculture, and Ford's tractors played a huge role in making this happen, has enabled one American farmer to feed on the order of 100 people, there are still enormous inefficiencies in much agricultural work. Videos and pictures of strawberry harvesting show, for example, large numbers of farm workers walking through strawberry fields to pick the berries, which they put into containers that they must then carry to collection points. They must do the two things that Henry Ford wrote that no job should ever require a worker to do; bend over and take more than one step in any direction. The job adds value only in the instant that the worker picks a berry and deposits it in a container, and intelligent farmers who cannot afford complex harvesting machines realize this. WGAL TV (2015) reports how Crop Care's PA 1600 Picking Assistant, a self-propelled trailer on which the worker lies face down, increases productivity enormously and requires far less physical effort. The crops appear to move beneath the worker as they would on a moving assembly line, and all he or she needs to do is reach down to get them. An overhead solar panel not only delivers power but shields the worker from the sun.

This also reinforces the key takeaway from this chapter that "Many see but few observe." Pictures and videos of farm workers bending and walking to harvest crops are available all over the Internet, and countless people complain about the low wages the farm workers receive. The problem is obvious in a fraction of a second to anybody who knows what to look for. Only when we realize what is wrong with the job's current design, do we have an incentive to improve it.

Opportunity Costs in Fruit Harvesting

The US Bureau of Labor Statistics reported that, in 2020, the median wage of agricultural workers was $13.89 an hour, which is far greater than the minimum wage. Ziprecruiter cites a somewhat lower wage for the specific occupation of fruit picking, which pays $25,356 a year as of May 2021. These are not abysmal wages but they are not good wages either, and a lot of fruit spoils because it is not harvested at the right time.

Recall that the job's productivity can be increased substantially with poles that allow workers to reach into trees without climbing ladders, and these can be purchased online for as little as $20.

This could conceivably improve productivity a few hundred percent, which makes it consistent with the removal of basic friction in a manual labor job. The Israeli company Tevel has however built drones that can identify fruits that are ready to pick and harvest them without bruising or other damage. A video of the machines in action is reminiscent of very large bees visiting the fruit trees where they select only ripe fruits and place them into containers. The product is therefore much less expensive for the consumer, the farmer makes more, and the people who operate and maintain the drones can receive much higher wages.

Tevel is far from the only company seeking to automate this labor-intensive job. FF Robotics has a YouTube video that shows a machine selecting apples and putting them into a conveyor belt. Courtney and Mullinax (2019) describe a "robotic arms race" to develop automated apple pickers, and when these become practical, the contest will again not be between high-wage and low-wage labor, but machine against machine with the machines in question being operated and maintained by high-wage workers.

Pay Attention to Materials and Energy

While access to cheap labor is the principal reason for offshoring, manufacturers can make an enormous amount of money by paying attention to materials and energy as well. *Any material or energy that enters a process and does not come out as a saleable product is waste that takes away directly from the bottom line.*

A drawback of the ISO 14001:2015 standard for environmental management systems is, in fact, its focus on "environmental aspects" or "[elements] of an organization's activities, products, or services that has or may have an impact on the environment." This is important because we must comply with environmental regulations and not discharge pollutants into the environment, but any material we throw away represents wasted money regardless of whether it is an environmental aspect.

Most wood can, for example, be discarded legally in any landfill, but Henry Ford recognized that he had paid for the wood and was therefore entitled to get value from the portion he could not use for wooden car components. Ford accordingly distilled the wood into saleable methyl alcohol and also the familiar Kingsford Charcoal that people use in barbecues even today. The proceeds brought in $12,000 a day in the money of the 1920s (Ford, 1926, 136). Dust from blast furnaces was also reused (Ford, 1926, 106–107); nothing was allowed to go to waste. Slag from steel manufacture is also relatively harmless, but Ford made his slag into cement and paving materials for resale (Ford, 1926, 98–99).

The Material and Energy Balance

The material and energy balance is an analytical technique similar to credit–debit accounting that is taught in sophomore chemical engineering. I know of no waste of material or energy that can hide from it because inputs and outputs must balance the same way credits and debits must balance.

The first step is to draw an analytical control surface around the process under consideration. This is synergistic with the Supplier, Input, Process, Output, and Customer (SIPOC) model. The next step is to account for all inputs including (1) items in the product's Bill of Materials (BOM), (2) consumables such as solvents and cutting fluids, and (3) energy. These must balance outputs in quantity and kind. Anything that goes in and does not come out as a saleable product is waste.

The ISO 50001:2018 standard for energy management systems adds a gap analysis, which puts into practice Emerson's comparison of what is (the current state) and what could be (the potential state). If for example a machining operation needs one kilowatt to transform parts, and it draws two kilowatts, we need to ask where half the energy is going because it is clearly not going into the parts. Hydraulic tools, for example, draw power even when they are not doing anything while electrical ones use electricity only when they are making something (Mayer, 2015).

Here is an example of materials. If 100 pounds of metal go into a machining operation, then 100 pounds must come out as product and chips. The chips might be recyclable, but they are not saleable (except perhaps to a recycler) and therefore represent waste. There are machining processes that grind 80 percent of the stock into scrap, i.e., 100 pounds of billets become 20 pounds of product and 80 pounds of scrap. Another way of saying this is that the process takes five billets, converts one into product, and sends the other four back (as chips) to be recycled. Henry Ford's workers took exception to any such machining waste because they recognized it as exactly that, and they looked for ways to reduce or eliminate it. Ford preferred to stamp or otherwise fabricate small parts and weld them together to avoid the waste involved in machining large castings.

The material and energy balance also forces process owners to pay attention to consumables, such as cutting fluids; they often take for granted. Expended cutting fluids are not salable and may even incur costs for disposal in compliance with environmental regulations. Manufacturers have discovered that cryogenic machining, e.g., with liquid nitrogen rather than a cutting fluid to avoid overheating the tool, not only eliminates the environmental waste and also occupational health and safety problems from contact with traditional coolants but also allows much faster machining (Okuma America, 2016). The key point here is to not take the consumables for granted, and the material and energy balance compels us to pay attention to them, but we need to look at the other advantages as well.

If cryogenic machining allows 50 percent faster operation (Richter, 2015, cites 52 percent faster for an application that involves titanium), then we effectively get three machine tools and three skilled operators for the price of two. This is simply an industrial application of Emerson's example in which having a warship that can fire its guns twice as rapidly as the competition is like having two ships and two crews for the price of one, and this simply reinforces what Emerson wrote about the struggle of efficiency versus inefficiency. If cryogenic cooling increases the machining speed by 400 percent for a given application (Sinkora, 2017 cites 40–400 percent), it's like having five machine tools and five skilled operators for the price of one. This means of course that the operators can be paid more, and the product will cost less not only because one worker is now five times as productive but also because the producer needs only one-fifth of the capital investment in machine tools. Yet another advantage includes longer tool life.

The material and energy balance can therefore be described in a single sentence: "everything that goes in must come out either as saleable product or waste." Remember that this includes consumables and energy as well as stock (such as metal billets) and items in the BOM. The material and energy balance also facilitates a material and energy review.

Material and Energy Review

ISO 50001:2018 clauses 6.3 through 6.6 can and should be extended to ISO 14001:2015 and expanded even further to encompass all forms of material waste as opposed to just environmental aspects. The material and energy balance is a powerful supporting tool for this.

- Clause 6.3 would become a material and energy review (as opposed to just an energy review) and require considerations of not only significant energy uses (SEUs) but also significant material uses (SMUs). The latter would encompass not only environmental aspects, but all purchased stock and consumables. Priority could be assigned on the basis of cost, which includes the cost of purchase and also the cost of disposal if the item is an environmental aspect. ISO 50001:2018's annex (A.6.3) adds that with regard to energy, the use of renewable energy does not represent an improvement. Even if we install an energy source that uses free wind power, hydroelectric power, or solar power, we can sell to the power grid any surplus so the waste of renewable energy is still waste.
- Clause 6.4 becomes material and energy performance indicators as opposed to just energy performance indicators. The baseline, as explained by Figure A.3 in ISO 50001:2018 in the standard, is the initial performance state that we seek to improve to a given target value. Emerson (1909, 17) used this principle when he compared the energy performance of a firefly to that of an incandescent lamp. The incandescent lamp's energy consumption was the baseline and the firefly's the achievable target.
- Clause 6.5 becomes a material and energy performance baseline. It should usually be possible to collect data that shows how much energy and how much stock and consumables are used in any process. Remember that inputs must balance outputs in both quantity and kind, and anything that goes in that does not come out as something we can sell is waste.
- Clause 6.6 becomes planning for collection of material and energy data. This also should be relatively straightforward on the process level.

The next step is to apply these basic concepts to some real-world case studies. The gaps between the current performance state and the potential performance state become obvious once we know what to look for.

Hunt the Coal Thief

Jagd auf Kohlenklau (Hunt for the Coal Thief) is a German board game of the Second World War vintage that encouraged energy conservation. It personified energy waste as a Coal Thief with a walrus-like mustache and rat-like front teeth. The German Wikipedia entry adds that he was created by Wilhelm Hohnhausen and modeled on the Butzemann or what we would call the bogeyman. Another word for this creature is boggart, which appeared in J.K. Rowling's Harry Potter series as an entity that would take the shape of whatever a person feared the most, and the related word "bogey" is an unknown radar contact that could be hostile. *It's essentially something we rarely see while it causes far more trouble than we realize.*

The depiction of the Coal Thief as a bogeyman or Butzemann is also consistent with the "Gremlins from the Kremlin" in the Second World War cartoon "Russian Rhapsody" in which the gremlins in question sabotaged German war equipment. They depicted themselves as "the little men who aren't there," and a search on this phrase comes up with crossword puzzle solutions that include "elves" and "gnomes." This brings a whole new meaning to the Coal Thief who, like the bogeymen, elves, and gnomes, steals energy but is very hard to find unless one knows what to look for. He's something you might glimpse out of the corner of your eye, and you know somehow that he's around because your home or business is using far more energy than it should, but you can't really see him. "Hunt for the Coal Thief" taught people what to look for so they could throw the Kohlenklau out of their homes and workplaces.

The Coal Thief also carries a sack of stolen coal on his back, which could be based on the bogeyman's counterpart known as the Sack Man who carries misbehaving children away in a sack; the German folk figure Krampus also appears in this role. The personification of waste (muda) in general as a bogeyman-like figure could be very useful to educate people on how to detect waste that hides in plain view, which was in fact the role of the Kohlenklau in Germany.

The Kohlenklau is still used by the Federation for Energy Consumers (Bund fur Energie Verbraucher, no date given). This website explains, "The coal theft campaign is probably the most extensive energy saving campaign that has ever been carried out." A magnified version of the board, with the pictures and lettering clearly visible, is available on the British Museum's website.

The board game highlights energy wastes and also ways to conserve energy. It even reminds people to use appliances in off-peak hours, when energy demand is lower. Sleeping with the lights or radio on wastes a lot of energy, noting that the radios of that era used vacuum tubes and drew a lot more power than modern ones, and the lights were incandescent rather than light-emitting diodes. This does not mean we want to waste electricity to run a solid state radio or LED light today, but the corresponding wastes were enormously greater in the 1940s.

The Federation for Energy Consumers' web page provides a lot of insights into energy gap analysis which could carry over into the application of ISO 50001:2018 today. "Kohlenklau's Rechenbuch," the Coal Thief's accounting book, quantifies seemingly minor energy wastes to show how they add up to enormous trouble. If, for example, electric clothing irons were left on all over Germany for one minute per day, this wasted 150,000 kilowatt-hours. This is believable if, and this is true even today, an electric iron can draw 1,000 watts. If nine million families left their irons on unnecessarily for one minute per day, then 9,000,000 times one kilowatt-minute divided by 60 minutes per hour = 150,000 kWh or 200,000 horsepower-hour.

The problem statement then asks how long a factory whose 50 lathes draw ten horsepower each could run on this wasted energy. As the factory requires 500 horsepower to run the lathes, 200,000 hp-hour divided by 500 hp = 400 hours or twenty 20-hour work days. This is consistent with Benjamin Franklin's observation, "Beware of little expenses; a small leak will sink a great ship," and it gets people to think about how seemingly minor wastes can accumulate into real trouble. Remember that money we don't spend on waste flows directly to the bottom line as profit.

Another series depicts the Coal Thief's helpers or accomplices. One picture features a business office with brightly lit chandeliers and the Coal Thief apparently thanking a director for leaving them on all night. If we look at some brightly lit city skylines well after business hours, the same lesson could be learned by the organizations in question. If you are a customer, employee, or investor in a business whose headquarters is emitting a lot of light, especially after work hours, that's where your money is going.

The Coal Thief marries a female amphibian creature known as the Wasserplansche, or "water splash" which seems to bring material wastes, at least of water, into the discussion. The Wasserplansche lodges in most homes and businesses where people must let a shower or faucet run while they wait for hot water to arrive from the water heater. Point of use of water heaters eliminate this waste, but their capital cost is often significant, in the hundreds of dollars range. Water waste also happens when we wash dishes by hand and allow the water to run down the drain. The waste of water is taken for granted but can add up to a substantial amount.

Another series about the Coal Thief's "shameful defeat" depicts how to stop him by sealing cracks and crevices through which heat can escape. Several images meanwhile depict what looks like two-story cookware, i.e., one pot or vessel on top of another so the one on top captures heat from the one beneath it. Double-boiler pots are used today, apparently to control the amount of heat that reaches the food in the upper pot. The temperature cannot exceed the condensation temperature of steam, 212 degrees Fahrenheit or 100 degrees Centigrade at sea level, while a gas or electric burner can. The same arrangement can however make the heat from one burner perform two jobs. The one on the bottom can boil soup and the steam can then, as opposed to being allowed to escape, heat something in the one on the top. A picture from the board game elaborates that water for washing can be heated in this manner, above a pot used for cooking. The principle of using what would otherwise be waste heat to perform a second task on far larger scales is meanwhile well known to mechanical and chemical engineers.

"The Coal Thief's Summer" shows how to spot leaks in stoves and ovens, apparently by holding a lit candle to see if a draft deflects the flame. Another encourages people to seal windows against leaks, and yet another shows how buildups of soot and ash above burners keep heat from reaching pots and kettles. This lesson carries over into coal-fired and oil-fired furnaces such as those used to heat homes and also to generate steam for turbines. Soot, ash, and other types of fouling prevent heat from the fire from reaching the water (in a boiler or hot water heater) or air (for a heating system) which means the wasted heat goes up the chimney instead.

The takeaway is that the Coal Thief series teaches people to observe where others would merely see. The wastes are obvious once we know what to look for.

Paint Parts, Not Air

Shigeo Shingo (Robinson, 1990, 101–102) presents a case study in which a pen manufacturer was getting complaints from nearby rice farmers about paint solvents that were getting into their water. Shingo took one look at the spray booth in which pen caps were being painted and asked whether the objective was to paint the parts or the air. The manager realized immediately that most of the paint was missing the pen caps and ending up in wastewater that found its way to the rice paddies. The operation, as designed, was not only causing an environmental problem, it was literally throwing money down the drain. Redesign of the operation to direct the spray onto the pen caps alone reduced material costs and fixed the environmental problem.

While a material and energy balance was not necessary to expose this obvious waste, it would have worked as follows. A certain amount of paint solvents and solids enter the spray booth every hour, and a certain number of pen caps (or other products) come out with a quantifiable coating of the solids. The difference represents (1) solvents lost to evaporation and (2) solvents and solids lost to overspray. Ford (1926, 67) described how a solvent for a coating operation was recovered by adsorption by charcoal followed by desorption with steam to recover roughly 90 percent of the evaporated solvent for reuse. This means that every gallon of solvent did the work of ten. Sinclair (1937, 61) wrote of this, "He perfected new processes—the very smoke which had once poured from his chimneys was now made into automobile parts."

Ford added that his operations also recovered 2,100 gallons of oil and cutting fluids (consumables) from steel shavings every day for reuse. This is the kind of thinking that reduces or eliminates pollution and saves money as well.

Fertilize Crops, Not Groundwater

Fertilizer runoff means that as explained by the Environmental Protection Agency (EPA, no date given), "Nutrient pollution has impacted many streams, rivers, lakes, bays and coastal waters for the past several decades, resulting in serious environmental and human health issues, and impacting the economy." This is self-explanatory, and we must also recognize that the unused nitrogen and phosphorus are materials thats the farmer has paid but for which the farmer has received no value.

If we apply a material balance to a traditional farm, we will see very quickly that a substantial portion, if not the majority, of the water and fertilizer that go in do not come out as saleable crops. This is of course how agriculture has worked for thousands of years, but "we've always done it that way" is not an excuse for using better methods as they become available. Little Leaf Farms in Massachusetts grows lettuce in huge greenhouses in which, from what I can see from their website:

1. The sole source of water is rain water from the roof. Note that, as water does not run into the ground, this is apparently adequate so the farm uses 90 percent less water than conventional farms.
2. The hydroponic environment means no nutrients are wasted, which eliminates both the cost of the waste and the environmental aspect.
3. Carbon dioxide from the natural gas heating system is used to promote growth.
4. Pesticides are not used, which eliminates yet another potential environmental aspect plus the cost of the pesticide. We can assume that the enclosure helps keep out harmful insects, and the website says natural predators such as ladybugs eat any that might enter.
5. The farm can operate year-round, which brings up yet another advantage to this kind of agriculture. Greenhouses and similar enclosures protect crops that might otherwise be destroyed by unseasonable frosts or droughts.

Upward Farms in Brooklyn, New York, is building another facility in Hanover Township Pennsylvania as of late 2022. This is expected to be the largest organic vertical farm in the world (Allabaugh, 2022). The farm uses a closed system in which waste from striped bass, which are themselves saleable as food, fertilize the crops. The website[6] claims that water use is only five percent that of a traditional farm while "Our breakthrough approach unlocks a soil microbiome that's a million times more dense, fertile, and productive than chemically treated soil or synthetic hydroponic indoor farms." Luck (2022) adds of Kalera's vertical farms that they require 97 percent less space than a traditional farm and can also operate year-round without having to worry about the weather and other issues that can damage harvests.

These examples and Shingo's "paint parts, not air" show that it can be extremely profitable to eliminate an environmental aspect and also the cost of materials that are purchased but do not add value.

Baptize Converts, Not Parts

The same principle carries over into electroplating and etching operations in which plating or etching solutions must be rinsed from the parts. We need to ask whether the objective is to baptize the parts with complete immersion or use a spray rinse to remove the chemical in a far more concentrated form. Concentrated wastes are generally easier to handle and treat than dilute ones. Processes in which the parts must be completely free of chemicals often use a preliminary spray rinse to remove the bulk of the chemical, and immersion only to get rid of any remaining traces. "Reducing Dragout with Spray Rinses" (Merit Partnership Pollution Prevention Project for Metal Finishers, no date given) elaborates further.

Counterflow rinses, in which water from the most dilute rinse tank is then used for the less dilute one, and/or a spray rinse, also conserve water. The water is made to work two or three times before it is sent for treatment and again in a relatively concentrated rather than a dilute form. This is how we get the Coal Thief's wife, the Wasserplansche (Water Splash), out of the process. "Modifying Tank Layouts to Improve Process Efficiency" (Merit Partnership Pollution Prevention Project for Metal Finishers, 1996) provides an example. The bottom line is that even if we can treat wastewater to dispose of it in accordance with environmental laws, it is cheaper to treat concentrated waste or, even better, not generate it in the first place.

Raise Meat, Not Animals

A material balance on beef cattle would show that for every six pounds of food consumed by the animal, one pound of salable beef is produced along with various "byproducts" (Beef Cattle Research Council, 2014). The exact ratio is not important, but the basic principle is. A growing animal will use a substantial amount of food for metabolism. The animal will also grow organs that are not edible, at least not by most people. It also costs money to transport an animal from a feedlot to a meat packing factory, and there are also veterinary costs. This is why meat is expensive, and its environmental impact is relatively high.

How can we have meat without animals, associated animal cruelty problems, and the occupational health and safety problems associated with the meat packing industry? We simply have to get rid of the paradigm, "This is how we have raised meat for thousands of years," and scientists and engineers have done exactly that by growing meat in bioreactors (de Sousa, 2021). The nutrients go into meat rather than metabolism, the meat packing factory and associated transportation are removed from the supply chain entirely, and there are no veterinary costs either.

Dye the Yarn, Not the Water

The traditional dyeing industry uses enormous quantities of water. Ranson (no date given) cites McDonough and Braungart (2002) to state, "on average, only 5% of the raw materials involved in the production and delivery processes is contained within a garment." Fabric of the World (2020, no author given) states that 60 kilograms of water are used per kilogram of yarn. This results, of course, in an environmental aspect plus the cost of the water itself. This is the Coal Thief's spouse, the Wasserplansche (Water Splash) at work. This reference adds, however, that the amount needed for synthetic fabrics is much lower than it is for natural fibers like cotton.

Twine Solutions in Israel has however found a way to avoid the use of water in the dying process with its Digital Selective Treatment (DST™) approach. The equipment deals with environmental aspects, of which there are far lower quantities to begin with because no water is used, by capturing excess ink and volatile organic compounds at the point of generation. Twine Solutions' TS-1800 digital thread and yarn dyeing system can be installed instead in the weaving mill's facility to deliver the obvious advantage of making the yarn in exactly the quantities needed, and exactly when they are needed, by the fabric-making process. The web page for the TS-1800 adds that the machine will make up to 1,800 meters (1.1 miles) of thread per hour but is also suitable for short runs. This means in turn that the textile mill can fill orders for large quantities or just a few square yards depending on what its own customers need.

Sell the Coal Chemicals, Don't Burn Them

There were no laws against sulfur dioxide emissions 100 years ago, but Henry Ford (1926, 106 and 175) recognized that it was definitely uneconomical to burn coal chemicals for heat. Coking of the coal @$5/ton yielded coal chemicals worth $12 which means Ford got coke for his blast furnaces for literally less than nothing. His process also removed sulfur from the coal for sale as ammonium sulfate fertilizer (Figure 5.14) which means that something that would have otherwise become acid rain was saleable as a useful product instead.

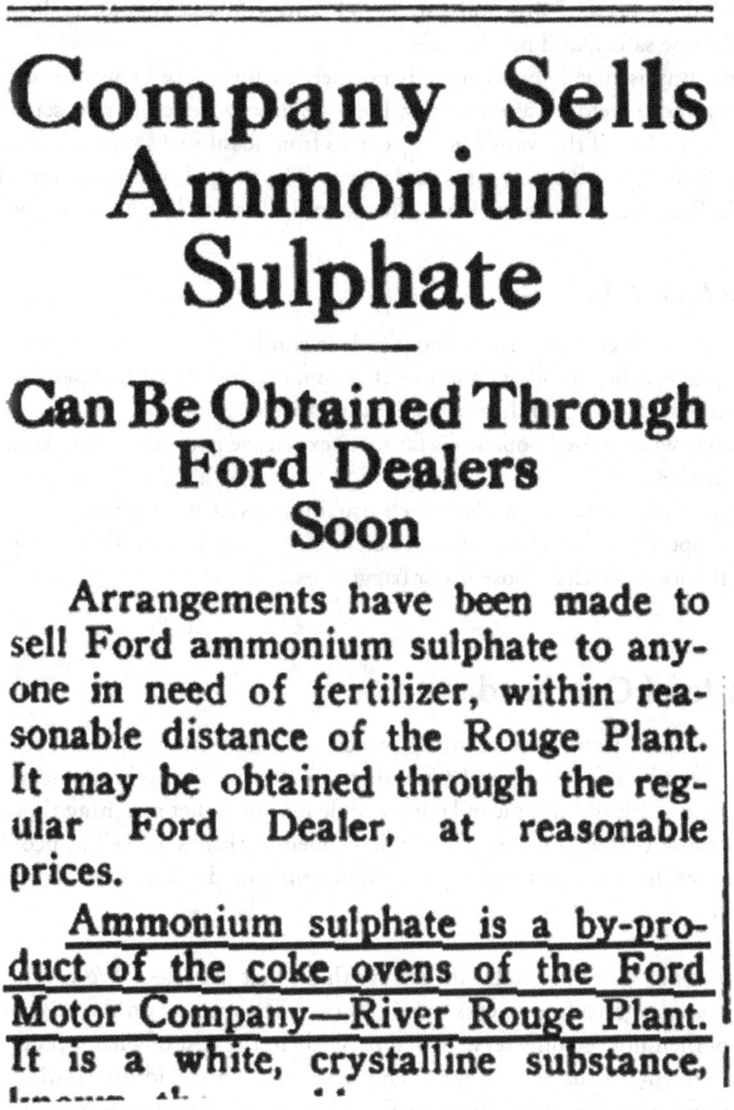

Company Sells Ammonium Sulphate

Can Be Obtained Through Ford Dealers Soon

Arrangements have been made to sell Ford ammonium sulphate to anyone in need of fertilizer, within reasonable distance of the Rouge Plant. It may be obtained through the regular Ford Dealer, at reasonable prices.

Ammonium sulphate is a by-product of the coke ovens of the Ford Motor Company—River Rouge Plant. It is a white, crystalline substance,

Figure 5.14 Ammonium Sulfate as Coke By-Product. ***Ford News*** **(1922, public domain due to age)**

Light the Streets, Not the Sky

Shigeo Shingo's admonition to paint parts rather than air applies to the use of light as well. Drake (2019) reports how, after an earthquake cut off power to Los Angeles, the emergency call system was overwhelmed by people who wanted to report a mysterious cloud in the sky. The cloud was the Milky Way, which they had never seen before due to light pollution. This wastes electricity and, if the power comes from fossil fuels, also results in unnecessary carbon emissions. Aerial views of Silicon Valley after dark are hardly encouraging here, and the same goes for other cities.

Light pollution also harms various animals, birds, and reptiles that do not know what to make of artificial light. Baby sea turtles die, for example, when light pollution guides them away from rather than toward the ocean. Directional street lights are available that focus the light where it is needed for nighttime safety, and nowhere else.

The key takeaway is that light where it is not needed for safety or work purposes represents wasted electricity and therefore wasted money. Light pollution over cities represents wasted money and also carbon emissions if the wasted energy comes from fossil fuel sources. Even if the energy is from renewable sources, it still costs money and may well be fungible via the power grid with energy from fossil fuels. Removal of the waste flows directly to the bottom line just like any other saving.

Ship Product, Not Air

Air is valuable only to divers, astronauts, and people in similar occupations where it is not available from the atmosphere. A box or other container that contains mostly air represents wasted shipping materials, wasted truck space and therefore shipping capacity, and/or wasted space on a store shelf. This is for example why popped popcorn is far more expensive per pound than kernels that can be popped in the kitchen.

The same principle applies to water, which usually costs at most pennies per gallon when it comes out of a pipe but otherwise costs a lot of money to transport. This is, for example, why bottled tea costs a lot more than loose tea or bagged tea.

Avoid Wasteful Overhead

Remember that a basic lean manufacturing principle is that anything that does not add value is waste. Henry Ford's rule (Ford and Crowther, 1922) was "everything and everybody must produce or get out." This did not mean laying people off but rather assigning those in non-value-adding jobs to those that added value. The rule applied to things as well as people because any capital expenditure that did not support production constituted waste. Ford (Ford and Crowther, 1922) added this,

> We will not put into our establishment anything that is useless. We will not put up elaborate buildings as monuments to our success. The interest on the investment and the cost of their upkeep only serve to add uselessly to the cost of what is produced—so these monuments of success are apt to end as tombs. A great administration building may be necessary. In me it arouses a suspicion that perhaps there is too much administration. We have never found a need for elaborate administration and would prefer to be advertised by our product than by where we make our product.

The world's first monuments of success, the Egyptian pyramids, were in fact tombs. Pharaohs planned well in advance because a big pyramid could require 20 or more years to build. The Great Pyramid at Giza required more than two million blocks, each of which weighed close to three tons. The cost of labor (even by unpaid workers under the robot or corvée system) had to have been astronomical, and they were not even valuable as tourist attractions at the time. Wolchover (2012) provides an estimate of $5 billion in the money of 2012 to build something like this with modern construction equipment, in comparison to $4 billion for the new World Trade Center. King Khufu therefore tied up a literal fortune in something he could not even use while he was alive, although our own era is grateful to him for providing one of the Seven Wonders of the Ancient World.

This is not, however, something that happened only several thousand years ago. Fancy office buildings and corporate headquarters cost money in terms of rent or, if owned by the company, money tied up in real estate, along with costs of maintenance and insurance. Ford wrote accurately that an asset is worth only what we can do with it. A house is a place to live without paying rent, and a corporate headquarters is a place to do business. It needs to be comfortable for the workers, and it must have all the necessary equipment, but it need not be opulent or in an expensive location like San Francisco or Los Angeles to perform its function.

Expensive Cities Add Costs

Another problem arises when companies locate their operations in costly cities, and/or states with high taxes and high costs of living. They have to pay higher wages and salaries to attract and keep workers. These higher wages are however not better wages than those paid in less expensive venues because they don't buy more actual value in terms of housing, groceries, and so on. Ford (Ford and Crowther, 1922) told us of this 100 years ago,

> And finally, the overhead expense of living or doing business in the great cities is becoming so large as to be unbearable. It places so great a tax upon life that there is no surplus over to live on. The politicians have found it easy to borrow money and they have borrowed to the limit. Within the last decade the expense of running every city in the country has tremendously increased. A good part of that expense is for interest upon money borrowed; the money has gone either into non-productive brick, stone, and mortar, or into necessities of city life, such as water supplies and sewage systems at far above a reasonable cost. The cost of maintaining these works, the cost of keeping in order great masses of people and traffic is greater than the advantages derived from community life. The modern city has been prodigal, it is to-day bankrupt, and to-morrow it will cease to be.

If we look again at the lessons of history, it is to be remembered that cities developed for exactly two reasons: places that could be defended against military attack and centers of commerce. The difficulty of attacking a walled city, castle, or, during the horse and musket era, a trace italienne fortress was often enough to put off most attackers. Even then, however, a good military engineer like France's Sébastien Le Prestre de Vauban and the Netherlands' Menno van Coehoorn (the inventor of the Coehorn Mortar) could capture one. The British proved at Copenhagen (1807) that even the weapons of the horse and musket era could go over a city's walls to destroy it. The Second World War proved that cities are no longer defensible against aircraft or ballistic missiles and are in fact high-density targets for them instead.

There was also a time when most major retail businesses and even manufacturing establishments were in cities, and "a trip to the big city" was a major event for families. The rise of suburban shopping malls displaced this function more than 50 years ago, and now we have home delivery by a wide variety of sellers. Business can be conducted from anywhere in the world via the Internet, so an office in the middle of "flyover country" can perform the same functions as one in a 100-story skyscraper in a big city. If people do not want to live in remote rural locations, there are plenty of small towns and suburbs with far lower real estate costs than those that prevail in big cities.

The COVID-19 epidemic meanwhile forced organizations to use online collaboration methods, and these were proven to work very well for most applications. This offers opportunities for the elimination of many facilities for physical presence entirely. Many classrooms, along with their capital costs and related commuting expenses, can be replaced with remote learning. International conferences can be held online to eliminate travel and lodging costs, along with risks related to (for example) bad weather that can interfere with air travel. If periodic face-to-face meetings are desired, rooms can be rented as needed by the day or even the hour.

Educate the American Consumer to Buy Value and Not Waste

There was a time when consumers in the United States understood the need to buy value and quality. There is a saying, and I do not know the origin, to the effect that a poor person can afford to buy only the best because he or she can't afford to buy it twice. The way to become rich and stay that way is to think like a poor person and do likewise, and this includes zero tolerance for shoddy goods regardless of whether they are made domestically or offshore.

As an example, I bought a GPS device for something on the order of $200, and it stopped working several years later because an internal component, and I recall that it was nothing more than a data storage device similar to a flash drive or SD drive, stopped working. There was no replacement part available and no way to even open the GPS to service it. The manufacturer told me it was out of warranty, which it was, but it would give me a 20 percent discount to buy a new one. I replied that I was not going to reward poor quality and nonexistent maintainability by buying another one and that the manufacturer was permanently disqualified from selling to me in the future. I had to buy a replacement, but it was from somebody else. My policy, and everybody else should adopt it as well, is that they will pay for something only once, and if they have to pay for it twice, the second time will not be from the business that sold the first one. Consumers also often have the option of writing negative online reviews for poor-quality goods or services.

Ford (Ford and Crowther, 1922) even applied this principle to motor vehicles, which we know do not last forever although the median vehicle age in the United States is now approaching twelve years.

> It is my ambition to have every piece of machinery, or other non-consumable product that I turn out, so strong and so well made that no one ought ever to have to buy a second one. A good machine of any kind ought to last as long as a good watch.

There was a time, perhaps, where some people felt they needed to have a new car every few years, but they are becoming rarer every year as vehicle prices rise. The person with the 15-year-old vehicle is, in fact, more likely to be a millionaire than the one with the 2-year-old vehicle because he or she has not sunk tens of thousands of dollars every few years into something that loses value the instant it is driven off the dealer's lot as pointed out by financial talk show host Dave Ramsey. Ramsey (2020) adds that valets with whom many millionaires entrust their cars are not impressed with the used Hondas, Camrys, or pickup trucks—until the valet gets the tip.

We have already seen that the price of any purchased item consists of (1) legitimate costs of capital, material, and labor, (2) a fair profit for the producer, and (3) waste. This book has considered so far wastes associated with production, such as waste motion. Other wastes may, however, be packed into the item's cost.

Don't Buy Indulgences or the Emperor's New Clothes

The basic issue is as follows. Our economy is full of "businesses" that are essentially nothing more than parasites that collect, and "earn" is emphatically not the right word, money from uninformed customers. The ancestors of these parasites included, for example, medieval indulgence sellers who took money to forgive people's sins (as if they had the power to do so) so they would spend less time in a conveniently invisible Purgatory from which, if it existed, dissatisfied customers could never complain.

Geoffrey Chaucer's "The Pardoner's Tale" depicts one such scoundrel who tells a story about three men who agree to destroy Death. They learn that Death may be found under a specific tree, where they find a pile of gold. Two conspire to murder the third after they send him into town to get transportation for the gold, and they do so when he returns only to die themselves when they drink the poisoned wine he purchased so he could have the gold for himself. Rudyard Kipling's "The King's Ankus" is a similar story in which a group of men find "Death" in the form of a jeweled elephant goad and kill one another so as to not have to share. Chaucer's Pardoner ends his tale with a sales pitch for indulgences he claims were issued by the Pope:

> Now, goode men, God forgeve yow your trespas,
> And ware yow fro the sinne of avarice [warn you from the sin of avarice].
> Myn holy pardoun may yow alle waryce [My holy pardon will cure you all],
> So that ye offre nobles or sterlinges [nobles and sterlings = coins]
> Or elles silver broches, spones, ringes [silver brooches, spoons and rings].

Hans Christian Andersen's "The Emperor's New Clothes" featured purported weavers who outfitted a monarch, at enormous expense, with "clothing" that was invisible to anybody who was incompetent. Neither the Emperor nor any of his subjects therefore dared to point out, when he appeared in public in his underwear, that he had been scammed. The twenty-first century's sellers of invisible and/or nonexistent products have simply become a bit more sophisticated. There are for example companies that sell water with "dissolved oxygen" for health benefits, which is not false advertising because oxygen is soluble in water. Other companies sell oxygen supplements to add to water and charge upward of $20. What they don't tell us, of course, is that tap water also contains dissolved oxygen which means they are selling something we can have for free. The digestive tract is not designed to extract dissolved oxygen anyway which makes even this aspect meaningless. Schwarcz (2021) points out that there is more oxygen in one breath of air than there is in a liter (a little more than a quart) of water. This section's key takeaway is, "Demand value for your money, and refuse to pay for waste."

Extended Warranties

Extended warranties are almost universally a total waste of the customer's money. This is because the manufacturer's warranty covers the period during which defects in manufacture are likely to make themselves known, after which the item will perform for a long time before it finally wears out. A simple model for this is known as the "bathtub curve" which depicts the rate of failure versus time. The initial rate of failure, which is covered by the manufacturer's warranty, is relatively high and is known as infant mortality. This results from manufacturing defects that escape from the factory. The bottom of the curve is both minimal and flat, and this represents the product's useful life. The extended warranty therefore covers the period during which the failure rate is at the absolute minimum. This may (in theory) be forever for electronic products with no moving parts. The last part of the curve, during which the failure rate increases, is wearout or old age. The extended warranty does not cover this period as shown in Figure 5.15.

Griffith (2020) says of extended warranties, "These companies aren't offering extended warranties out of altruism. They're doing it because service plans make crazy amounts of money." He

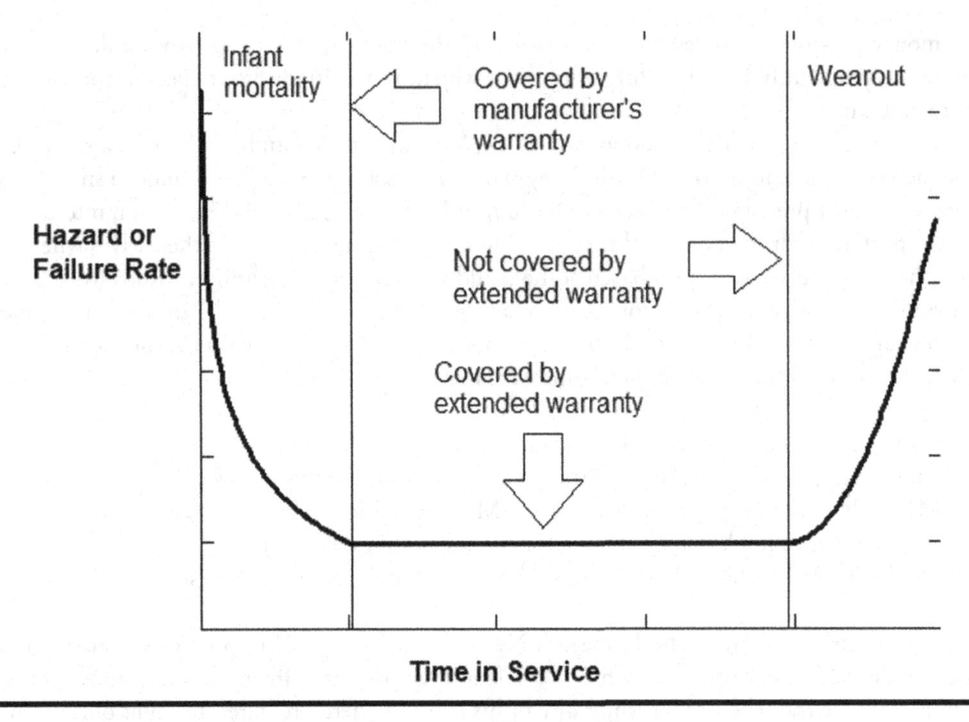

Figure 5.15 Why Extended Warranties Are Usually Worthless

adds that extended warranties make $40 billion a year for their sellers, which means consumers may as well be throwing $40 billion a year into the nearest dumpster.

Ramsey (2021) adds of extended vehicle warranties sold by car dealers, "The National Auto Dealers Association says that the average car dealer is losing money on the sale of each new car, so tacking on the warranty is crucial for them to make money." He adds that roughly half of what the customer pays goes to the sales representative's commission, which underscores the status of these extended warranties as total waste. Remember that the manufacturer's warranty covers the entire vehicle for (usually) at least a year, and the power train for much longer.

Car dealers also try to get something for nothing by selling the modern counterparts of indulgences such as "rust proofing" (cars are manufactured to be relatively rust-proof), "fabric protection" (a perfunctory spray from a cheap aerosol can for which the dealer might add $100), "paint protection," and of course extended warranties (Luthi, 2021). Financial talk show host Dave Ramsey meanwhile refers to car leases as "fleeces" and with good reason. ONeal (2021) adds, "Leasing also happens to be the most expensive way to drive a car" and explains why. A consumer zero-tolerance attitude toward waste, and insistence on value for money, will go a long way toward increasing personal wealth and reshoring value-adding jobs.

Advertising

Advertising beyond the amount necessary (such as the cost of a website and other keyword-searchable media) to inform potential customers about products also adds no value. "Pull" advertising is similar to pull production because buyers who want something, or need a solution to a particular problem, will search the Internet and/or catalogs to which they subscribe it to find sellers who can meet their needs. "Push" advertising, such as that in newspapers, magazines, radio, and television, is directed to people who more likely than not do not need what is being sold because, if they did, they would have probably found it online or in catalogs to which they subscribe. Some "push" advertising may be necessary to let the public know of one's existence, but, on the other hand, ads that are repeated over and over serve only to tell savvy customers that the cost must be carried by the price tag of the item.

The same applies if ads appear every month in one's mailbox, as I have seen for a particular credit card. There is a dumpster within easy walking distance of my mailbox, and that is where these go, unopened, before they even get through my door. If they are going into everybody's mailbox (and probably from there into the dumpster), then one or more stakeholders must be carrying the cost of this waste. I have also regretted donating money to certain charities that began to send me monthly newsletters and even small-value gifts like address labels, notepads, and greeting cards. While I can use some of them, they obviously represent money not being spent on the group's charitable mission and therefore waste.

Some cars carry a "New Car Regional Advertising Fee" which can be as high as two percent of the manufacturer's suggested retail price (Bernstein, 2015), or $600 for a $30,000 car. The dealer may as well mark the $600 as "waste," and customers should simply refuse to pay it. The question is as to whether the dealer will then back away from a $30,000 sale because the customer refuses to pay for waste. My position as a customer would be simply,

> This fee adds no value to the car and I will not pay it. If you insist on payment as a condition of closing the deal, there are 1 and 2 year old used cars that have almost as much life in them as this new one, their sticker price is much lower because a car loses a good part of its value the instant I drive it off your lot, and they do not carry this waste in their price tag.

This argument becomes far more persuasive near the end of the model year when, if the dealership cannot sell the vehicle in time, it will have to knock several thousands rather than $600 or so off the price when it becomes last year's model. Waiting until the end of the model year might be a good idea anyway because last year's model still has the same life in it if it has not been driven.

"Push" advertising may even be a holdover from the time when people could not search the Internet for what they wanted. There was a time when, if somebody wanted a car or a house, he or she had to read the local paper's classified ads to find one. Newspapers still carry this kind of advertising, but it is far less relevant now than it was even 30 years ago.

Suppose for example I see car or real estate advertisements in my local paper and decide, for some reason, I want a new car or house. I don't need to limit my choices to what I see in the paid ads. If I do an Internet search on "homes for sale" and my zip code, the first result is Realtor .com which returned 101 homes when I tried it. If I do the same with "cars for sale," I get links to Cars.com, Carfax, Auto Trader, and Edmunds. Even if the ads are successful in creating a need I didn't know I had, all they do is suggest that I look on the Internet to find what is quite likely to be a better deal.

Soft drink and other beverage companies have meanwhile paid millions of dollars for television ads during the Super Bowl, and these ads can cost upward of $5 million for 30 seconds. It certainly gets the brand's name out there but, when people go to the supermarket or a warehouse store to actually buy soft drinks or beer, they will see competitors' products as well, and some of these may have lower prices.

Ads for insurance also raise questions. If we assume that all insurance companies use similar actuarial methods to predict (for example) how many insured properties will incur losses in any given year, the cost of covering these losses along with profits and administrative costs will be roughly similar for everybody. These costs, along with advertising costs, must be carried by insurance premiums and the advertising costs do not add value for the purchasers.

The popularity of ad-blocking add-ons for Internet browsers, such as Adblock Plus, FB Purity, and uBlock Origin, should meanwhile make it clear to sellers that the ads for which they must pay, and then add the cost to their products and services, are not only of no value to customers; they are of negative value. *If your ads are being blocked, it should be obvious that the viewers regard them as non-value-adding clutter and are unlikely to buy whatever it is you are trying to advertise.* A company called BlockAdBlock offers a script with which websites can intrude a pop-up over their material if they detect an ad blocker, which looks like a good way to get viewers to abandon the website or install an extension to disable BlockAdBlock. There are also extensions that will disable the pop-ups themselves. Web domains whose sole purpose is to serve ads can be banned from a computer entirely via the HOSTS file. This is done simply by adding lines with "127.0.0.1 [url of ad serving domain]" which makes your computer unable to connect with the domains in question. Many digital video recorders (DVRs) meanwhile offer features that remove commercials from recorded television programs, which means the advertiser may well be paying for something many people don't even see.

Most people understand, however, that advertising does help pay for websites and will not block noninvasive banner ads (this is Adblock Plus's default setting) even if they don't pay much attention to them. Ads that invade the middle of the content and pop-up ads are however far more likely to be blocked and/or their sources disabled via the HOSTS file which makes them a complete waste of money for the advertiser. If the website charges advertisers per impression (the number of times the ad is displayed) as opposed to per interaction or click-through, this could be a warning that advertising money is going to waste.

Celebrity Endorsements and Brand Names

A celebrity's paid product endorsement is a good reason to not buy the product in question because the endorsement cost is built into the price, and it adds no value for the consumer. An athletic jersey with a celebrity's name on it is for example "imported" and sells for more than $100. Similar "imported" jerseys, minus the celebrity's name, can be purchased for $20 to $30. Chadha (2017) reports that a survey of millennials by Roth Capital Partners found that "78%, either had a negative view of celebrity endorsements, or were indifferent to the practice with regard to making a purchase." This should come as no surprise because celebrities can and will accept money to endorse almost anything, as long as the product or service is not so unsavory that they do not want to associate their names with it.

The same principle goes for a fancy brand name unless of course, as was the practice 50 or so years ago, the brand went with value and quality. A brand name today is as likely as not to just be a sales tool with no added value. Most readers know meanwhile that the store brand of a detergent or medication is chemically identical to the brand name but costs a lot less, which means people who buy the brand name are paying a premium for a label. It is noteworthy, for example, that you can buy certain "premium" ice cream brands only in highly uneconomical one-pint containers while the store brand comes in three-pint containers for the same price or even a lower one. Some big box stores sell gallon quantities at even lower prices per unit. If the seller won't let you buy its so-called premium brand in economical quantities, it's charging you for labels rather than product.

Another example of poor value for your money consists of attractive dog treats that look like gourmet cupcakes, albeit with dog-safe ingredients. Their appearance is designed to appeal to humans rather than dogs, just as the appearance of some expensive fishing lures is meant to appeal to fishermen rather than fish. Humans are very receptive to the appearance of their food as shown by pictures in restaurant menus, while dogs are not. The dog smells something to determine whether it is edible, and this determination happens very quickly. If it is edible, the dog's next agenda is to get it into his or her stomach as rapidly as possible. Ordinary peanut butter (without dangerous additives like Xylitol) and ordinary dog treats will do just as well as "gourmet" dog treats. I have seen, in fact, one "gourmet" dog treat that is not only expensive but also unhealthy for dogs due to its sugar content.

When the American consumer wakes up to the fact celebrity endorsements and/or fancy designer labels add no value to shoddy goods made by cheap offshore labor, businesses will have to sell value instead of celebrity endorsements and fancy designer labels. This is what they had to do in the first part of the twentieth century when the few people who could afford to pay for waste were unwilling to do so.

Cryptocurrencies

A celebutante is somebody who is essentially famous for being famous, and cryptocurrencies are similarly valuable only because people think they are valuable, as they once thought that Dutch tulip bulbs and dot-com stocks were valuable. It is easy to envision Hendrik Pot's "Wagon of Fools" featuring cryptocurrency logos instead of the tulip bulbs that were popular at the time (Figure 5.16).

Figure 5.16 Wagon of Fools. Pot, Hendrik Gerritsz. 1637. Dutch tulip speculators abandon their value-adding trades to ride the wagon into the sea ("Wagon of Fools," public domain due to age.)

Another painting of that era by Jan Brueghel the Younger depicted tulip speculators as brainless monkeys headed toward debtors' court and the grave, noting that some people paid as much for a tulip as they might for a house.

Matters have reached the point where the electrical power necessary to "mine" cryptocurrencies by performing complex computation exceeds the amount necessary to mine rare earths as well as platinum, gold, and copper (Hern, 2018, Nov. 5). This reference says it requires 19 megajoules to "mine" one dollar worth of Bitcoin, versus nine for a dollar's worth of rare earths and five for a dollar's worth of gold. To put this in perspective, a kilowatt equals 1,000 joules per second so a kilowatt-hour (3,600 seconds) equals 3.6 megajoules. This means it costs 5.2 kilowatt-hours to create a dollar's worth of Bitcoin, as opposed to less than two to obtain a dollar's worth of gold which is actually useful in dentistry, jewelry, electronics manufacture, and aerospace. Slightly less than three kilowatt-hours for a dollar's worth of rare earths that are actually useful for, among other things, rechargeable batteries, lasers, and DVD players, also are a very good investment.

Hern (2018, Jan. 17) adds that Bitcoin mining generates as much carbon dioxide as a million transatlantic flights. Aratani (2021) adds, "Electricity needed to mine bitcoin is more than used by 'entire countries.'" This means that people are buying enormous quantities of electrical energy to generate a commodity whose value depends entirely on the perception that it is valuable which means that, as was the case for tulip bulbs and dot-com stocks, a devastating cryptocurrency crash is only a matter of time. Cryptocurrencies did, in fact, take a major dip in May 2021 which suggests that people are waking up to the fact that paying for computers to burn enormous quantities of electricity to perform trillions of blockchain calculations is a bad idea. Bitcoin was, as of mid-October 2022, about 70 percent off its high of roughly $64,400 as of November 12 2021. Anything that is valuable just because people think it is valuable, like tulip bulbs, stocks in 1929, dot-com stocks in 2001, and houses in 2007–2008, is a disaster waiting to happen and cryptocurrencies are no exception.

Private versus Public Universities

Student debt is another major issue, and many people incur it by going to pricy private universities instead of state-subsidized public ones. There is little practical difference in the quality of education provided by either source. An argument could even be made that freshman and sophomore courses taught at community colleges are of higher quality because the instructors are hired solely to teach, as opposed to performance of research. Incentives such as "publish or perish" and the need to get research grants mean that many professors at research institutions give less attention to teaching which, at least at the freshman level, sometimes takes place in auditoriums where students receive little individual attention except from teaching assistants whose highest priority also is research.

The issue is best summed up as follows: "What do you call somebody who graduated from an inexpensive medical school instead of a famous one, and then passed the licensing exam? 'Doctor.'" Dave Ramsey (Leonhardt, 2019) cited meanwhile the case of a student who chose an expensive university because the campus was pretty and would be likely to graduate with $40,000 in debt as a result. The purpose of college is to gain job skills with which the graduate can earn a living, and it makes no sense to spend six figures at an Ivy League or similar institution when a state university, or two years at a community college followed by two at a state university, will deliver the same education.

Summary

The first chapter showed that the quantity theory of money, which relates the money supply and velocity of money to prices and quantities of transactions, shows that higher productivity enables (1) lower prices, (2) higher wages, and (3) greater employment simultaneously. This chapter has covered off-the-shelf approaches to productivity improvement, some of which have been in use for more than 100 years. Enormous quantities of waste often hide in plain view, but they are usually easy to remove once we know they are there. This chapter has hopefully provided enough examples and case studies to equip the reader to recognize these often-elusive wastes.

It is also important, however, for consumers to recognize and demand value for their money. Brand names, celebrity endorsements, extended warranties, and advertising costs do not add value or utility to the product or service so consumers should refuse to pay for these things. This is especially true when sellers put price tags suitable for American-made products on cheap offshore goods.

Conclusion

The first chapter addressed the issue of inflation, which is the easily foreseeable result of too much money chasing too few goods. The Federal Reserve's tools with which to address this have both limitations and unpleasant side effects, as investors know from the first three quarters of 2022. The United States also faces an ever-increasing deficit that will, if left unchecked, spiral out of control when we must start to borrow money to pay the interest on the principal. Higher productivity, however, enables simultaneous high wages, high profits, and low prices which counteracts inflation and also increases taxable economic activity. The United States has had these tools for more than a hundred years thanks to efficiency pioneers like Frank Gilbreth, Harrington Emerson, Frederick Winslow Taylor, and Henry Ford. They are ours to use again any time we feel like doing so, and the past few years have proven clearly that the time is now.

The second chapter underscored the enormous danger of loss of manufacturing capability, which is a universal indicator of national decline. There have been no exceptions. Manufacturing is also the backbone of military power as proven during the Second World War, which makes it very bad judgment to transfer it to a hostile foreign power like the People's Republic of China. Manufacturing also has the power, however, to remove the root causes of war by making cooperation and trade far more profitable than conflict.

The third chapter showed that the People's Republic of China is a dangerous geopolitical rival that has menaced other countries in the region with violence and also has a very long track record of selling American individual, corporate, and government customers substandard and counterfeit products that can put life and safety at risk. The American consumer should not tolerate made-in-America prices for shoddy and often dangerous PRC-made goods. Even responsible and reputable offshore suppliers in Europe, Japan, Taiwan, South Korea, and so on are however subject to force majeure which introduces unnecessary risks into any supply chain. Remember that the total cost of ownership, or total cost of use, of a seemingly inexpensive product can include enormous costs that are not reflected by the ostensible purchase price.

The fourth chapter demonstrated that cheap labor is often symptomatic not of high profits and low prices, but rather low profits and excessive prices because it gives waste a perfect place to hide. History has proven that intelligently managed high-wage labor outperforms cheap labor and even unpaid labor, such as that which still exists in the form of robot or corvée. Financial metrics can be enormously dysfunctional because, when we ask the wrong questions or use the wrong performance measurements, we will get the wrong answers.

The fourth chapter also discredited Luddism, the delusion that automation and higher productivity will put people out of work. The Luddites have been wrong for well over a thousand years, and they will always be wrong unless short-sighted managers do something, such as laying people off in response to productivity improvements, to prove them right.

The first four chapters showed therefore why it is vital to reshore manufacturing and increase its productivity. The fifth chapter provided the details as to how to do this. The ancient Greeks taught the world how to think around problems rather than trying to solve them head-on with brute force, which includes large quantities of cheap labor instead of a handful of Frederick Winslow Taylor's intelligently managed and properly equipped high-priced workers.

Friction and opportunity costs come from astronomical wastes that are built into many if not most jobs, and they can remain there for decades or even longer because they are taken for granted. Anybody who knows what to look for, and the fifth chapter provided numerous case studies to illustrate the principles, will recognize these wastes on sight. Off-the-shelf methods to remove friction include standard work, for which multiple references are available.

Any material or energy input that does not become a salable process output is also waste, but no material or energy waste can hide from a material and energy balance. Outputs must balance inputs in both kind and quantity, and it is easy enough to differentiate the saleable outputs from the waste. Wasteful overhead, such as expensive office buildings and also expensive places to live, adds cost but no value to the organization's product or service. The American consumer can meanwhile play a central role by demanding value for his or her money.

Notes

1. Memnon was portrayed by Peter Cushing, a classical horror actor who was later famous as Grand Moff Tarkin in the original Star Wars movie, in the 1956 movie starring Richard Burton
2. "Frederick Taylor's Pig Iron Carrying Experiment tests the efficiency of workers based on how many rest periods they receive," created in the 1910s.
3. Voltaire's *Candide* features "Bulgarian" recruiters who might be stand-ins for Prussians (noting their blue uniforms) who persuade Candide to take some money and drink their King's health, whereupon he discovers to his amazement that he has joined their army and must now obey their orders. It is conceivable that Voltaire disguised his Prussians as Bulgarians as he was on friendly terms with Frederick the Great and did not want to depict the King of Prussia as a common crimper.
4. Lord (2022) reports that Russians are now performing Internet searches on how to break their own arms to avoid conscription into Vladimir Putin's army.
5. 29 CFR 1926.1053(b)(22) states, "An employee shall not carry any object or load that could cause the employee to lose balance and fall."
6. https://upwardfarms.com/our-story

Bibliography

Agence France-Presse. 2021. "Taiwan's Worst Drought in Decades Deepens Chip Shortage Jitters." *Industry Week*, April 20. https://www.industryweek.com/supply-chain/article/21161812/taiwans-worst-drought-in-decades-deepens-chip-shortage-jitters.

Ainsworth, W. Harrison Esq (editor). 1852. "Firearms." *The New Monthly Magazine and Humorist*, volume 94.

Allabaugh, Denise. 2022. "Upward Farms Celebrates Completion of New Farming Facility in Hanover Twp." *Citizens Voice*, September 16, 2022.

Altieri, M.A., and Koohafkan, P. 2004. "Globally Important Ingenious Agricultural Heritage Systems (GIAHS): Extent, Significance, and Implications for Development." April 10, 2019. www.fao.org/docrep/015/ap021e/ap021e.pdf

Anthony, Robert, and Reece, James. 1983. *Accounting Text and Cases*, 7th ed. Homewood, IL: Richard D. Irwin Inc.

APICS Staff. 2011. "Majority of Businesses Experienced Supply Chain Failure." (no longer available online).

Aratani, Lauren. 2021. "Electricity Needed to Mine Bitcoin is More than Used by 'Entire Countries'." *The Guardian*. https://www.theguardian.com/technology/2021/feb/27/bitcoin-mining-electricity-use-environmental-impact.

Aristotle. 350 BCE. *Politics*. Translation in Works of Aristotle: Politica, by B. Jowett, Oeconomica, by E.S. Forster, Atheniensium respublica, by Sir F. G. Kenyon. 1921. Oxford: Clarendon Press. https://books.google.com/books?id=mL1MAAAAYAAJ.

Arnold, Horace Lucien, and Faurote, Fay Leone. 1915. *Ford Methods and the Ford Shops*. New York: The Engineering Magazine. Reprinted 1998, North Stratford, NH: Ayer Company Publishers, Inc.

Arrington, Benjamin T. n.d. "Industry and Economy during the Civil War." National Park Service. https://www.nps.gov/articles/industry-and-economy-during-the-civil-war.htm.

Automotive Industry Action Group/Verband der Automobilindustrie (AIAG/VDA). 2019. *Failure Mode Effects Analysis Handbook*. Southfield, MI: AIAG/VDA.

Automotive News. 2011. "Chrysler, GM Cut Output at 6 Plants due to Storm-Related Carpet Shortage." https://www.autonews.com/article/20110914/OEM/309149706/chrysler-gm-cut-output-at-6-plants-due-to-storm-related-carpet-shortage.

Barboza, David. 2008. "In Chinese Factories, Lost Fingers and Low Pay." *New York Times*. https://www.nytimes.com/2008/01/05/business/worldbusiness/05sweatshop.html.

Barleen, Steven. 2018. *Manufacturing*. https://www.encyclopedia.com/social-sciences-and-law/economics-business-and-labor/businesses-and-occupations/manufacturing.

Basset, William R. 1919. *When the Workmen Help You Manage*. New York: The Century Co.

Batha, Emma. 2020. "China Accused of Forcing 570,000 People to Pick Cotton in Xinjiang." Thomson Reuters Foundation, https://www.reuters.com/article/china-cotton-forced-labour-trfn/china-accused-of-forcing-570000-people-to-pick-cotton-in-xinjiang-idUSKBN28P2CM.

BBC. 2020. "UK Business 'Must Wake Up' to China's Uighur Cotton Slaves." December 16, 2020. https://www.bbc.com/news/business-55319797.

Beef Cattle Research Council. 2014. "How Much Feed and Water are Used to Make a Pound of Beef?" https://www.beefresearch.ca/blog/cattle-feed-water-use-2014/.

Benson, Allan L. 1923. *The New Henry Ford*. New York: Funk and Wagnalls, p. 256.

Bernstein, Alex. 2015. "What are New Car Regional Advertising Fees?" *CarsDirect*. https://www.carsdirect.com/car-pricing/what-are-regional-advertising-fees.

Bickford, Ethan. 2012. "War Time Counterfeits." *Paper Money Guaranty*. https://www.pmgnotes.com/news/article/2572/War-Time-Counterfeits/.

Biden, President Joseph. 2022. "Remarks by President Biden on the Chips and Science Act at IBM Poughkeepsie." October 6, 2022. https://www.whitehouse.gov/briefing-room/speeches-remarks/2022/10/06/remarks-by-president-biden-on-the-chips-and-science-act-at-ibm-poughkeepsie/.

Blois, Matt. 2022. "The US Solar Industry has a Supply Problem." *Chemical & Engineering News*, September 18, 2022. https://cendigitalmagazine.acs.org/2022/09/18/the-us-solar-industry-has-a-supply-problem-2/content.html.

Boggan, Steve. 2015. "Gold Rush California was Much More Expensive than Today's Tech-Boom California." *Smithsonian Magazine*. https://www.smithsonianmag.com/history/gold-rush-california-was-much-more-expensive-todays-dot-com-boom-california-180956788/.

Boot, Max. 2006. *War Made New: Technology, Warfare, and the Course of History, 1500 to Today*. New York: Gotham Books.

British Museum. n.d. "Jagd auf Kohlenklau (Hunt for the Coal Thief)." https://www.britishmuseum.org/collection/object/P_2004-1231-15.

Buncombe, Andrew. 2020. "US and China in War of Words as Beijing Threatens to Halt Supply of Medicine Amid Coronavirus Crisis." *The Independent*. https://www.independent.co.uk/news/world/americas/us-politics/coronavirus-china-us-drugs-trump-rubio-china-virus-xinhua-hell-epidemic-a9400811.html.

Buxbaum, Peter. 2018. "$5 Billion in US Scrap Exports in Jeopardy." *American Journal of Transportation*. https://www.ajot.com/premium/ajot-5-billion-in-us-scrap-exports-in-jeopardy.

Buxbaum, Peter. 2020. "China Trade and COVID-19: A One-Two Punch Hitting US Hardwood Producers." *American Journal of Transportation*. https://ajot.com/premium/ajot-china-trade-and-covid-19-a-one-two-punch-hitting-us-hardwood-producers.

Carey, Chris. 2019. "The Puzzles of Thermopylae." *History Today*. https://www.historytoday.com/miscellanies/puzzles-thermopylae.

Carlson, Lowell. 1974. "The History of Cotton Strippers." *Gas Engine Magazine*. https://www.gasenginemagazine.com/farm-life/the-history-of-cotton-strippers/.

Carr, Lawrence. 1987. *Lectures in Management Control Systems*. Poughkeepsie, NY: Union College. (Quotes are from my handwritten notes and may not be verbatim.)

CBS News. 2015. "Petco Pulls Pet Treats from China Suspected of Killing, Sickening Thousands." https://www.cbsnews.com/news/petco-pulls-chinese-pet-treats-suspected-of-killing-sickening-thousands/.

Chadha, Rahul. 2017. "Millennials are Wary of Celebrity Endorsements." *Insider Intelligence*. https://www.insiderintelligence.com/content/millennials-are-wary-of-celebrity-endorsements.

Chakraborty, Barnini. 2020. "China Hints at Denying Americans Life-Saving Coronavirus Drugs." *Fox News*. https://www.foxnews.com/world/chinese-deny-americans-coronavirus-drugs.

Chemical Engineering Progress. 2022. "Solar Tower Produces Jet Fuel Sustainably." September 2022, p. 4.

Clausewitz, Carl von. 1976. *On War*. Translated by M. Howard and P. Paret. Princeton, NJ: Princeton University Press.

Collins. 2022. "How the US Economy Lost Its Independence, and Workers Their Livelihoods." *Industry Week*. https://www.industryweek.com/the-economy/trade/article/21245772/how-the-us-economy-lost-its-independence-and-workers-their-livelihood.

Conniff, Richard. 2011. "What the Luddites Really Fought Against." *Smithsonian Magazine*. https://www.smithsonianmag.com/history/what-the-luddites-really-fought-against-264412/?page=2.

Copp, Tara, and Baldor, Lolita C. 2022. "Pentagon Says China is Top Threat." *Associated Press, The Citizens Voice*, October 28, 2022.

Courtney, Ross, and Mullinax, T.J.. 2019. "Washington Orchards Host Robotic Arms Race — Video" *GoodFruit Grower*. https://www.goodfruit.com/washington-orchards-host-robotic-arms-race/.

Covey, Stephen R. 1991. *Principle Centered Leadership*. New York: Simon & Schuster.

Crouch, Gregory. 1989. "Safety Threat Seen: Counterfeits Now Nuts, Bolts Issue." *Los Angeles Times*, January 27. https://www.latimes.com/archives/la-xpm-1989-01-27-mn-1479-story.html.

Crow, Carl. 1943. *The Great American Customer*. New York: Editions for the Armed Services, Inc.

Customs and Border Protection. 2021. "CBP Seizes Counterfeit N95 Masks." https://www.cbp.gov/newsroom/local-media-release/cbp-seizes-counterfeit-n95-masks.

Daniel, Will. 2022. "Billionaire Investor Carl Icahn Warns "The Worst is Yet to Come" for Investors and Compares U.S. Inflation to the Fall of the Roman Empire." *Yahoo Finance*. https://finance.yahoo.com/news/billionaire-investor-carl-icahn-warns-165840843.html.

DeLuzio, Mark. 2021. "Standard Work." Lean Horizons Consulting, webinar presented May 25.

Department of Energy, Health, Safety and Security. 2007. *Suspect/Counterfeit Items Awareness Training*. https://www.energy.gov/sites/default/files/2014/06/f16/SCI_Training_Manual.pdf.

Desjardins, Jeff. 2016. "This Infographic Shows How Currency Debasement Contributed to the Fall of Rome." *Business Insider*. https://www.businessinsider.com/how-currency-debasement-contributed-to-fall-of-rome-2016-2.

de Sousa, Agnieszka. 2021. "Meat Grown in Israeli Bioreactors Is Coming to American Diners." *Bloomberg*. https://www.bloomberg.com/news/articles/2021-06-23/meat-grown-in-bioreactors-is-coming-to-american-diners-next-year.

D'Innocenzo, Anne. 2021. "Toy Makers Race to Get Products on Shelves Amid Supply Clogs." *Tulsa World*, October 6, 2021. https://tulsaworld.com/business/ap/toy-makers-race-to-get-products-on-shelves-amid-supply-clogs/article_df35b9b8-783c-57fd-aa26-24acb4bbcab8.html.

Dickens, Charles. 1843. *A Christmas Carol*. London: Chapman & Hall.

Dimitri, Carolyn, Effland, Anne, and Conklin, Neilson. 2005. "The 20th Century Transformation of U.S. Agriculture and Farm Policy." U.S. Department of Agriculture. https://www.ers.usda.gov/webdocs/publications/44197/13566_eib3_1_.pdf.

Drake, Nadia. 2019. "Our Nights are Getting Brighter, and Earth is Paying the Price." *National Geographic*, April 3.

Ebel, Roland. 2019. "Chinampas: An Urban Farming Model of the Aztecs and a Potential Solution for Modern Megalopolis." *HortTechnology*, 30(1). https://journals.ashs.org/horttech/view/journals/horttech/30/1/article-p13.xml.

Emerson, Harrington. 1909. *Efficiency as a Basis for Operation and Wages*. New York: The Engineering Magazine.

Emerson, Harrington. 1924. *The Twelve Principles of Efficiency*, 6th edition. New York: The Engineering Magazine. Original copyright, 1911 by John R. Dunlap.

Environmental Protection Agency. n.d. "Nutrient Pollution; The Issue." https://www.epa.gov/nutrientpollution/issue.

Ericksen, Paul, and McKinney, Eamon. 2021. "With China Policy in Tatters, OEMs Best Expect More Shortages." *Industry Week*, April 14. https://www.industryweek.com/supply-chain/supply-chain-initiative/article/21161289/with-china-policy-in-tatters-oems-best-expect-more-shortages.

Fabric of the World. 2020. "Water Wastage & Contamination Caused by Irresponsible Fabric Dyeing Processes." https://www.fabricoftheworld.com/post/water-wastage-contamination-caused-by-irresponsible-fabric-dyeing-processes.

Fast, Larry. 2017. "Are 'Concrete Heads' Wrecking Your Lean Manufacturing Efforts?" *Industry Week*. https://www.industryweek.com/operations/continuous-improvement/article/22009541/are-concrete-heads-wrecking-your-lean-manufacturing-efforts.

FDA. 2009. "Melamine Pet Food Recall - Frequently Asked Questions." https://www.fda.gov/animal-veterinary/recalls-withdrawals/melamine-pet-food-recall-frequently-asked-questions.

Federation for Energy Consumers (Bund fur Energie Verbraucher). n.d. "The Coal Thief." https://www.energieverbraucher.de/de/der-kohlenklau__1446/.

Ferry, Jeff. 2020. "It's Time to Rebuild Domestic Drug Production in the US, for Both Health and Economic Reasons." *Industry Week*, March 17, 2020. https://www.industryweek.com/the-economy/article/21126380/its-time-to-rebuild-domestic-drug-production-in-the-us-for-both-health-and-economic-reasons.

Fisher, Irving. 1922. *The Purchasing Power of Money*. New York: Macmillan.

Ford, Henry. 1922. *Ford Ideals: From "Mr. Ford's Page."* Dearborn: The Dearborn Publishing Company.
Ford, Henry, and Crowther, Samuel. 1922. *My Life and Work.* New York: Doubleday, Page & Company. https://www.gutenberg.org/cache/epub/7213/pg7213.html.
Ford, Henry, and Crowther, Samuel. 1926. *Today and Tomorrow.* New York: Doubleday, Page & Company (Reprint available from Productivity Press, 1988).
Ford, Henry, and Crowther, Samuel. 1930. *Moving Forward.* New York: Doubleday, Doran, & Company.
Ford News, 1922. Dearborn MI
Franklin, Benjamin. 1758. "The Way to Wealth." http://www.swarthmore.edu/SocSci/bdorsey1/41docs/52-fra.html.
Fraser, George MacDonald. 1969. *Flashman.* New York: Penguin Group.
Freedberg, Sydney. 2016. "Excalibur Goes to Sea: Raytheon Smart Artillery Shoots Back." *Breaking Defense.* https://breakingdefense.com/2016/01/excalibur-goes-to-sea-raytheon-smart-artillery-shoots-back/.
Gallagher, John. 2002. "Back on Board." *Journal of Commerce Online.* https://www.joc.com/maritime-news/back-board_20021201.html.
Ganzel, Bill. 2007. "Cotton Harvesting." *Wessels Living History Farm.* https://livinghistoryfarm.org/farminginthe50s/machines_15.html.
Gilbreth, Frank Bunker. 1911. *Motion Study.* New York: D. Van Nostrand Company.
Glasberg, Davita. 1989. *The Power of Collective Purse Strings; The Effects of Bank Hegemony on Corporations and the State.* University of California Press. Chapter Two—"W. T. Grant: The Social Construction of Bankruptcy." https://publishing.cdlib.org/ucpressebooks/view?docId=ft4x0nb2jj&chunk.id=d0e1075&toc.depth=1&toc.id=d0e1075&brand=ucpress.
Glatter, Robert. 2020. "Almost 70% of Chinese KN95 Masks Don't Meet Minimum Safety Standards." *Forbes.* https://www.forbes.com/sites/robertglatter/2020/09/25/almost-70-of-chinese-kn95-masks-dont-meet-minimum-safety-standards/.
Goldratt, Eliyahu, and Cox, Jeff. 1992. *The Goal*, 2nd revised ed. Croton-on-Hudson, NY: North River Press.
Goodman, Peter S. 2021. "'It's Not Sustainable': What America's Port Crisis Looks Like Up Close." *New York Times*, October 10. https://www.nytimes.com/2021/10/11/business/supply-chain-crisis-savannah-port.html.
Government Accountability Office. 2022. "Larger Federal Deficits & Higher Interests Rates Point to the Need for Urgent Action." https://www.gao.gov/blog/larger-federal-deficits-higher-interests-rates-point-need-urgent-action.
Griffith, Eric. 2020. "Here's Why an Extended Warranty on Electronics Is a Waste of Money." *PCMag*, July 27. https://www.pcmag.com/how-to/heres-why-an-extended-warranty-on-electronics-is-a-waste-of-money.
Gross, Daniel. 2002. "How the West Coast Port Shutdown Could Ruin Christmas." *Slate*, October 2. https://slate.com/business/2002/10/how-the-west-coast-port-shutdown-could-ruin-christmas.html.
Grossman, Dave. 1996. *On Killing.* New York: Back Bay Books.
Hughes, Daniel. 1993. *Moltke on the Art of War: Selected Writings.* New York: Presidio Press.
Halpin, J.F. 1966. *Zero Defects.* New York: McGraw-Hill.
Hanson, Victor Davis. 1989. *The Western Way of War.* New York: Oxford University Press.
Hanson, Victor Davis. 2001. *Carnage and Culture.* New York: Anchor Books.
Hanson, Victor Davis. 2020. "Is America a Roaring Giant or Crying Baby?" *Chicago Tribune*, April 8. https://www.chicagotribune.com/opinion/commentary/ct-opinion-coronavirus-america-roaring-giant-hanson-20200408-yo6i4sdaxvhqnbavnwni3p57bq-story.html.
Hern, Alex. 2018. "Bitcoin's Energy Usage is Huge – We Can't Afford to Ignore it." *The Guardian*, January 17. https://www.theguardian.com/technology/2018/jan/17/bitcoin-electricity-usage-huge-climate-cryptocurrency.
Hern, Alex. 2018. "Energy Cost of 'Mining' Bitcoin More than Twice that of Copper or Gold." *The Guardian*, November 5. https://www.theguardian.com/technology/2018/nov/05/energy-cost-of-mining-bitcoin-more-than-twice-that-of-copper-or-gold.
Hollings, Alex. 2018. "Counterfeit Air Power: Meet China's Copycat Air Force." *Popular Mechanics.* https://www.popularmechanics.com/military/aviation/g23303922/china-copycat-air-force/.

Indiana University of Pennsylvania. n.d. "LED Lighting Benefits." https://www.iup.edu/energymanagement/howto/led-lighting-benefits/.

Industry Week Staff. 2021. "Toyota Announces 40% Worldwide Production Cut Due Next Month." August 19, 2021. https://www.industryweek.com/supply-chain/article/21172830/toyota-announces-40-worldwide-production-cut-due-next-month.

Inskeep, Steve. 2022. "China's Ambassador to the U.S. Warns of 'Military Conflict' over Taiwan." *NPR*. https://www.npr.org/2022/01/28/1076246311/chinas-ambassador-to-the-u-s-warns-of-military-conflict-over-taiwan.

The Iron Age. 1897. "The Uehling Casting Machine." *The Iron Age*, April 22, pp. 12–14.

Jacobs, W.W. 1902. *The Monkey's Paw*. New York: Harper's Monthly.

James, Peter, and Thorpe, Nick. 1994. *Ancient Inventions*. New York: Ballentine Books.

Japsen, Bruce. 2011. "$625,000 Judgment against Baxter in 2007 Blood-Thinner Death Case." *Chicago Tribune*. https://www.chicagotribune.com/business/ct-xpm-2011-06-09-ct-biz-0610-baxter-heparin-20110609-story.html.

Jebb, Richard Claverhouse. 1893. *Sophocles*, 3rd ed., the seven plays, text. http://classics.mit.edu/Sophocles/antigone.html.

Juran, Joseph. 1995. *A History of Managing for Quality*. Hinsdale, IL: Irwin Professional Publishing.

Kay, Grace. 2021. "Truckers at Backlogged Ports Say They've Waited in Miles-Long Lines for up to 8 Hours without Pay." *Business Insider*. https://www.businessinsider.com/truckers-wait-outside-backlogged-ports-8-hours-without-pay-2021-11.

Keegan, John. 1993. *A History of Warfare*. New York: Vintage Books.

Koh, Ann. 2021. "Shipping Containers Plunge Overboard as Supply Race Raises Risks." *Bloomberg*. https://finance.yahoo.com/news/shipping-containers-plunge-overboard-supply-210000175.html.

Krisher, Tom. 2021. "Parts Shortage Will Keep Auto Prices Sky-High." *Citizens Voice*, September 6.

Kuo, Lily, and Wintour, Patrick. 2020. "Hong Kong: China Threatens Retaliation against UK for Offer to Hongkongers." *The Guardian*. https://www.theguardian.com/world/2020/jul/02/china-could-prevent-hongkongers-moving-to-uk-says-dominic-raab.

Lao Tzu. 1963. Translation by D.C. Lau. *Tao Te Ching*. London: Penguin Books.

Lash, Herbert, Sanyal, Shreyashi, and Biswas, Ankika. 2022. "Wall Street Tumbles as Jobs Report Cements Harsh Rate Hike Outlook." *Reuters*. https://finance.yahoo.com/news/us-stocks-wall-street-tumbles-183242146.html.

Lawson, Thomas P.E.. 2017. "Bloomsburg Flood Management Project." Keystone PSPE meeting, April 9.

Lee, Yen Nee. 2021. "2 Charts Show How Much the World Depends on Taiwan for Semiconductors." *CNBC*. https://www.cnbc.com/2021/03/16/2-charts-show-how-much-the-world-depends-on-taiwan-for-semiconductors.html.

Leo, Geoff. 2020. "Health Canada Issues Recall of Some KN95 Masks Made in China." *CBC*. https://www.cbc.ca/news/canada/saskatchewan/health-canada-issues-recall-of-some-kn95-masks-made-in-china-1.5568734.

Leon-Porfilla, M. 1992. *The Aztec Image of Self and Society, An Introduction to Nahua Culture*. Salt Lake City, UT: University of Utah Press.

Leonhardt, Megan. 2019. "Money Expert Dave Ramsey Tells Students: Skip the "Dream" College and Go to School Where You Can Afford." *CNBC*. https://www.cnbc.com/2019/10/07/dave-ramsey-tells-students-go-to-school-where-you-can-afford.html.

Levin, Carl (U.S. Senator). 2011. "The Committee's Investigation into Counterfeit Electronic Parts in The Department of Defense Supply Chain." U.S. Senate Hearing. https://www.govinfo.gov/content/pkg/CHRG-112shrg72702/html/CHRG-112shrg72702.htm.

Levinson, William. 2017. "Culture: A Decisive Competitive Advantage." *Quality Digest*. https://www.qualitydigest.com/inside/lean-column/culture-decisive-competitive-advantage-100317.html.

Lewis, Katie. 2009. "China's Counterfeit Medicine Trade Booming." *Canadian Medical Association Journal*, 181(10): E237–E238. https://www.cmaj.ca/content/181/10/E237.

Lieven, Anatol. 2020. "Stay Calm About China. Beijing's Ambitions Shouldn't Be Treated as an Existential Threat to the United States." Foreignpolicy.com. https://foreignpolicy.com/2020/08/26/china-existential-threat-united-states-xi-jinping/.

Linderman, Frank B. 2002. *Plenty-Coups: Chief of the Crows*, 2nd ed. Lincoln, NE: University of Nebraska Press.

Little, Robert. 2001. "U.S. Merchant Fleet Sails toward Oblivion." *Baltimore Sun*, August 6. https://www.baltimoresun.com/bal-te.bz.sealift06aug06-story.html.

Long, Gideon. 2015. "Humberstone: A Chilean Ghost Town's English Past." *BBC News*. https://www.bbc.com/news/world-latin-america-31090757.

Longworth, Philip. 1965. *The Art of Victory. The Life and Achievements of Field-Marshal Suvorov*. New York: Holt, Rinehart, and Winston.

Lord, Debbie. 2022. "Why are Russians Googling 'How to Break Your Arm at Home'?" *Cox Media Group National Content Desk*. https://www.wftv.com/news/trending/why-are-russians-googling-how-break-your-arm-home/Z4ZQC7LZJ5EB3IM2PVQBRPCK4M/.

Luck, Marissa. 2022. "In Texas, High-Tech Vertical Farms are Transforming How Houstonians Get Their Greens." *Houston Chronicle*. https://www.houstonchronicle.com/business/real-estate/article/In-Texas-high-tech-vertical-farms-are-17479146.php.

Luthi, Ben. 2021. "5 Dealer Options to Skip When Buying a Car." *Bankrate*. https://www.bankrate.com/loans/auto-loans/dealer-options-to-skip-when-buying-a-car/.

Luvaas, Jay. 1966. *Frederick the Great on the Art of War*. New York: The Free Press.

Madison, James. 1794. "Commercial Discrimination." https://founders.archives.gov/documents/Madison/01-15-02-0120.

Mahan, Alfred Thayer. 1890. *The Influence of Sea Power Upon History 1660–1783*. Boston, MA: Little, Brown and Company. https://www.gutenberg.org/files/13529/13529-h/13529-h.htm.

Mannix, Daniel P. 1958. *The Way of the Gladiator*. New York: iBooks Inc (2001 reprint).

Mayer, Matthew. 2015. "Flipping the Switch on E-Powered Machine Tools." *Manufacturing Engineering*, October 2015.

Mayo, Juan (presenter, online webinar). 2022. *Hexavalent Chromium Safety*. Pennsylvania Training for Health and Safety (PATHS), September 27, 2022.

McDonough, William, and Braungart, Michael. 2002. *Cradle to Cradle: Remaking the Way We Make Things*. New York: North Point Press. (imprint of Farrar, Straus and Giroux).

Merit Partnership Pollution Prevention Project for Metal Finishers. n.d. "Reducing Dragout with Spray Rinses." https://archive.epa.gov/region9/waste/archive/web/pdf/metal-spray.pdf.

Merit Partnership Pollution Prevention Project for Metal Finishers. 1996. "Modifying Tank Layouts to Improve Process Efficiency." https://www.nmfrc.org/pdf/other/0256.pdf.pdf .

Michon, Kathleen. n.d. "Heparin Recall and Litigation; Problems with Heparin Have Prompted an FDA Recall and Numerous Lawsuits over the Anticoagulant Drug." *Nolo*. https://www.nolo.com/legal-encyclopedia/heparin-recall-litigation-32951.html.

Miley, Timothy. 2015. "Animal Product Injuries from Defective Products." *Avvo*. https://www.avvo.com/legal-guides/ugc/animal-product-injuries-from-defective-products.

Minter, Steve. 2013. "How Risky is the Defense Supply Chain?" *Industry Week*, June 2013, p. 64.

Mishkin, Frederic S. 1986. *The Economics of Money, Banking, and Financial Markets*. Boston: Little, Brown & Company.

Mitchell, Henry. 1895–1896. "Viscount Ferdinand De Lesseps." *Proceedings of the American Academy of Arts and Sciences*, 31: 370–384.

Moore, Kelly. 2011. "Living because of Linen? Professor's Project Takes a Shot at Ancient Armor." University of Wisconsin. https://news.uwgb.edu/featured/05/19/linen-professors-project-ancient-armor/.

Morello, Sandra. 2020. "Timber Processor Alarmed over "Massive Amount" of Softwood Exported to China." *ABC South East SA*, September 1. https://www.abc.net.au/news/2020-09-02/softwood-resource-exported-to-china-amid-log-supply-insecurity/12617994.

Morgan, David. 2008. "The True Price of Auto Labor Costs." *CBS News*. https://www.cbsnews.com/news/the-true-price-of-auto-labor-costs/.

Moser, Harry. 2015. "Overview of Reshoring Benefits, Costs, Opportunities, and Risk Considerations." Talking Freight Webinar. https://www.fhwa.dot.gov/Planning/freight_planning/talking_freight/july_2015/index.cfm.

Moser, Harry. 2021. "Total Cost of Ownership." *Industrial Heating.* https://www.industrialheating.com/articles/96277-total-cost-of-ownership.

"Moving Assembly Line Debuts at Ford Factory". *History.* https://www.history.com/this-day-in-history/moving-assembly-line-at-ford.

Myers, Andrew. 2022. "Stanford Engineers Create a Catalyst that can Turn Carbon Dioxide into Gasoline 1,000 Times More Efficiently." *Stanford News.* https://news.stanford.edu/2022/02/09/turning-carbon-dioxide-gasoline-efficiently/.

Nash-Hoff, Michele. 2012. "Senate Report Reveals Extent of Chinese Counterfeit Parts in Defense Industry." *Industry Week*, May 31. https://www.industryweek.com/the-economy/article/21957101/senate-report-reveals-extent-of-chinese-counterfeit-parts-in-defense-industry.

National Geographic. n.d. "Mansa Musa (Musa I of Mali)." https://education.nationalgeographic.org/resource/mansa-musa-musa-i-mali/.

National Institute of Standards and Technology (NIST). n.d. "NIST Reveals How Tiny Rivets Doomed a Titanic Vessel." https://www.nist.gov/nist-time-capsule/nist-beneath-waves/nist-reveals-how-tiny-rivets-doomed-titanic-vessel.

National Iron and Steel Publishing Company. 1905. *Steel and Iron.* Volume 76, Issue 22.

Nicholson, Marcy. 2021. "Sawmills are Selling Boards Faster than They Can Cut Them." *American Journal of Transportation.* https://ajot.com/news/sawmills-are-selling-boards-faster-than-they-can-cut-them.

Nikkei Asia (Nikkei Staff Writers). 2022. "Japanese Companies Explore How to Go 'Zero-China' Amid Tensions." https://asia.nikkei.com/Spotlight/Supply-Chain/Japanese-companies-explore-how-to-go-zero-China-amid-tensions.

Norwood, Edwin P. 1931. *Ford: Men and Methods.* Garden City, NY: Doubleday, Doran & Company Inc.

Occupational Health and Safety Administration. n.d. "Transcript for the OSHA Training Video Entitled Counterfeit & Altered Respirators: The Importance of Checking for NIOSH Certification." https://www.osha.gov/video/respiratory-protection/niosh/transcript.

Ohno, Taiichi. 1988. *Toyota Production System: Beyond Large-Scale Production.* Portland, OR: Productivity Press.

Okuma America. 2016. "Little-Known Facts About Cryogenic Machining." https://www.okuma.com/blog/blog-facts-cryogenic-machining.

ONeal, Anthony. 2022. "How Does Leasing a Car Work?" https://www.ramseysolutions.com/debt/how-does-a-car-lease-work.

"On the Pre-Modern Waterfront." 2002. *Wall Street Journal*, October 3, 2002. http://www.wsj.com/articles/SB1033612069728300833.

Ordonez, Victor. 2021. "US Bans All Cotton and Tomato Products from Xinjiang over Slave Labor." *ABC News.* https://abcnews.go.com/International/us-bans-cotton-tomato-products-xinjiang-slave-labor/story?id=75226217.

Palmer, R.R., and Colton, Joel. 1971. *A History of the Modern World.* New York: Alfred A. Knopf.

Panchak, Patricia. 2014. "Did Finance Gut Manufacturing?" *Industry Week*, June 9, 2014.

Peplow, Mark. 2022. "The Race to Upcycle CO2 into Fuels, Concrete and More." *Nature.* https://www.nature.com/articles/d41586-022-00807-y.

Peters, Tom. 1987. *Thriving on Chaos.* New York: Harper and Row.

Phillips, Charles, and Jones, David. 2015. *The Complete Illustrated History of the Aztec and Maya.* Leicester, UK: Anness Publishing.

Picchi, Aimee. 2019. "Almost Half of All Americans Work in Low-Wage Jobs." *CBS News*, December 2. https://www.cbsnews.com/news/minimum-wage-2019-almost-half-of-all-americans-work-in-low-wage-jobs/.

Powell, Bill. 2011. "The Global Supply Chain: So Very Fragile." *Fortune.* http://fortune.com/2011/12/12/the-global-supply-chain-so-very-fragile/.

Productivity Press Development Team. 2002. *Standard Work for the Shopfloor.* New York: Productivity Press.

Quality Progress. 2021. "Chip Shortage Causes Automakers to Hit Pause." June 2021, p. 11.

Radio Free Europe/Radio Liberty. 2013. "Uzbek Cotton-Picking Claims Eighth Victim." https://www.rferl.org/a/25145827.html.

Rajkumar Agro Engineers Pvt Ltd. YouTube Video. https://youtu.be/ESdL2JqpaWw.

Ramsey, Dave. 2020. "What Secret Millionaires Don't Tell You - Dave Ramsey Rant." https://youtu.be/HHw8o73m1ro.

Ramsey, Dave. 2021. "Just Say "No" to Extended Warranties." Ramsey Solutions. https://www.ramseysolutions.com/debt/just-say-no-to-extended-warranties.

Ramsey, Dave. 2021. "HELOC: What Is a Home Equity Line of Credit?" Ramsey Solutions. https://www.ramseysolutions.com/real-estate/home-equity-line-of-credit.

Ranson, Beth. n.d. "The True Cost of Colour: The Impact of Textile Dyes on Water Systems." Fashion Revolution. https://www.fashionrevolution.org/the-true-cost-of-colour-the-impact-of-textile-dyes-on-water-systems/.

Rare Historical Photos. n.d. "Using Banknotes as Wallpaper during German Hyperinflation, 1923." https://rarehistoricalphotos.com/banknotes-german-hyperinflation-1923/.

Regalado, Francesca. 2020. "China Poses a "Security Threat" to Japan, Taro Kono Says." *Nikkei Asia*. https://asia.nikkei.com/Politics/International-relations/China-poses-a-security-threat-to-Japan-Taro-Kono-says.

Reuters. 2022. "Ford to Halt Production Next Week at Flat Rock Plant on Chips Shortage." April 29, 2022. https://www.reuters.com/business/ford-halt-production-next-week-flat-rock-plant-chips-shortage-2022-04-29/.

Reuters. 2022. "Chinese Nationalist Commentator Deletes Pelosi Tweet after Twitter Blocks Account." July 30, 2022. https://www.reuters.com/world/china/chinese-nationalist-commentator-deletes-pelosi-tweet-after-twitter-blocks-2022-07-30/.

Reynolds, Isaac, Sanders, Allen, and Hillman, Douglas. 1983. *Principles of Accounting*, 3rd ed. New York: Dryden Press.

Richter, Alan. 2015. "Cryogenic Machining Systems can Extend Tool Life and Reduce Cycle Times." Cutting Tool Engineering. https://www.ctemag.com/news/articles/cryogenic-machining-systems-can-extend-tool-life-and-reduce-cycle-times.

Robinson, Alan (editor). 1990. *Modern Approaches to Manufacturing Improvement: The Shingo System*. Portland, OR: Productivity Press.

Roser, Christoph. 2015. "230 Years of Interchangeable Parts – A Brief History." https://www.allaboutlean.com/230-years-interchangeability/.

Royal, James. 2022. "Popularity of Cryptocurrency Plummets Among Millennials." *Bankrate*. https://www.bankrate.com/investing/cryptocurrency-popularity-declines-among-millennials-survey-shows/.

Rugaber, Christopher. 2022. "Slower US Job Gain in August Could Aid Fed's Inflation Fight." *Associated Press*, September 2. https://www.wsaz.com/2022/09/02/fed-is-hoping-august-hiring-report-will-show-slowdown/.

Rugaber, Christopher. 2022. "Fed Will Indicate Interest "Pain" at Meeting." *Associated Press*, September 20. https://www.wsaz.com/2022/09/02/fed-is-hoping-august-hiring-report-will-show-slowdown/ .

Sainato, Michael. 2021. "'This Used to Be a Great Job': US Truckers Driven Down by Long Hours and Low Pay." *The Guardian*. https://www.theguardian.com/business/2021/dec/27/us-truck-drivers-economy-pay-conditions.

Savitz, Eric. 2011. "Japan Quake Knocked Out 25% Of Global Silicon Wafer Production." *Forbes*. http://www.forbes.com/sites/ericsavitz/2011/03/21/japan-quake-knocked-out-25-of-global-silicon-wafer-production/.

Sawyer, Ralph D., with Sawyer, Mei-chün. 1993. *The Seven Military Classics of Ancient China*. Boulder, CO: Westview Press.

SBAM Shooting. 2016. "Fast Shooting the 'Old Way' with the Austrian M1784 Musket." *YouTube*. https://youtu.be/hohpriqPgEg.

Schwarcz, Joe. 2021. "Is There Any Point in Drinking Oxygenated Water?" McGill Office for Science and Society. https://www.mcgill.ca/oss/article/health-and-nutrition-you-asked/there-any-point-drinking-oxygenated-water.

Sekora, Michael. 2014. "Reviving U.S. Manufacturing Is the Wrong Goal to Set to Improve the Economy." *Forbes*, February 4. http://www.forbes.com/sites/beltway/2014/02/04/why-reviving-u-s-manufacturing-is-the-wrong-goal-to-set-to-improve-the-economy/.

Sheckley, Robert. 1968. *The People Trap*. New York: Dell Books.

Shepardson, David. 2022. "China Opposes Semiconductor Bill Because It Will Give U.S. Advantage - U.S. Commerce Chief." Reuters, May 11, 2022. https://www.reuters.com/technology/china-opposes-semiconductor-bill-because-it-will-give-us-advantage-us-commerce-2022-05-11/.

Shih, Gerry. 2020. "China Threatens Invasion of Taiwan in New Video Showing Military Might." *Washington Post*. https://www.washingtonpost.com/world/asia_pacific/china-taiwan-invasion-military-exercise/2020/10/12/291f5d86-0c58-11eb-b404-8d1e675ec701_story.html.

Shirouzu, Norihiko. 2001. "Job One: Ford Has Big Problem Beyond Tire Mess: Making Quality Cars." *Wall Street Journal*, May 25, 2001, A1 and A6.

Simon, Scott. 2005. "A Priest's Early Quest to Create a Bulletproof Vest." *NPR*. https://www.npr.org/2005/04/30/4625858/a-priests-early-quest-to-create-a-bulletproof-vest.

Simpson, Ian. 2012. "Flood of Fake Chinese Parts in US Military Gear – Report." *Reuters*. https://www.reuters.com/article/usa-defense-counterfeit/flood-of-fake-chinese-parts-in-us-military-gear-report-idUSL1E8GMMCV20120522.

Sinclair, Upton. 1937. *The Flivver King*. Second printing, 1987. Chicago, IL: Charles H. Kerr Publishing Company.

Sinkora, Ed. 2017. "Cryo Cooling Improves Machining of Super-Hard Materials, Gummy Polymers." Society of Manufacturing Engineers. https://www.sme.org/technologies/articles/2017/november/cryo-cooling-improves-machining-of-super-hard-materials-gummy-polymers/.

Smialek, Jeanna, and Ngo, Madeleine. 2021. "Supply Chain Woes Still a Threat." *Citizens Voice*, August 24, 2021. https://www.nytimes.com/2021/08/23/business/economy/supply-chain-bottlenecks-coronavirus-inflation.html.

Sorensen, Charles E., with Samuel T. Williamson. 1956. *My Forty Years with Ford*. New York: W. W. Norton & Company Inc.

Steadman, Jim. 2014. "John Deere Updates Round Module Building Cotton Picker for 2015." *Cotton Grower*. https://www.cottongrower.com/cotton-production/john-deere-updates-round-module-building-cotton-picker-for-2015.

Stern, Boris. 1939. *Labor Productivity in the Boot and Shoe Industry*. Philadelphia, PA: Works Progress Administration.

Steuben, Friedrich Wilhelm Ludolf Gerhard Augustin, Baron von. 1779. *Regulations for the Order and Discipline of the Troops of the United States*. Philadelphia, PA: Styner and Cist.

Stillwell, Blake. 2019. "6 Simple Reasons the North Won the Civil War." *Business Insider*. https://www.businessinsider.com/6-simple-reasons-the-union-north-won-the-civil-war-2019-11.

Student Action with Farmworkers. 2012. "Facts About North Carolina Farmworkers." https://saf-unite.org/wp-content/uploads/2020/12/nc-farmworkers-2012.pdf.

Stuelpnagel, T. R. 1993. "Déjà Vu: TQM Returns to Detroit and Elsewhere." *Quality Progress*, September: 91–95.

Stupak, Bart (Member of Congress). 2008. "The Heparin Disaster: Chinese Counterfeits and American Failures." U.S. Government Printing Office. https://www.govinfo.gov/content/pkg/CHRG-110hhrg53183/html/CHRG-110hhrg53183.htm.

Taylor, Frederick Winslow. 1911. *The Principles of Scientific Management*. New York: Harper Brothers. 1998 republication by Dover Publications, Inc., Mineola, NY. https://www.gutenberg.org/cache/epub/6435/pg6435.html.

Taylor, Frederick Winslow. 1911. *Shop Management*. New York: Harper & Brothers Publishers.

Techspex. 2014. "Conveyor Roller Systems–Since Recorded History." https://www.techspex.com/blog/post/conveyor-roller-systemssince-recorded-history

Tennenhouse, Erica. 2019. "These Fishermen-Helping Dolphins Have Their Own Culture." *National Geographic*. https://www.nationalgeographic.com/animals/article/dolphins-fishermen-brazil-culture.

The System Company. 1911. *How Scientific Management is Applied*. London: A. W. Shaw Company Ltd.

Times Sunday Special. 1898. "American Gunners Surprise the World." June 19, p. 25.

Toyota Times Global. 2020. "How it Happened: Toyota Production System Leads to 100-Fold Increase in Protective Gown Production." *YouTube*. https://youtu.be/knMN22_jQHI.

United Nations Conference on Trade and Development (UNCTAD). 2020. "Merchant Fleet." https://stats.unctad.org/handbook/MaritimeTransport/MerchantFleet.html.

UPI. 1988. "New Theory on Defeat of Spanish Armada" Cites Historian Geoffrey Parker's "Why the Armada Failed" from MHQ." *The Quarterly Journal of Military History*. https://www.upi.com/Archives/1988/10/02/New-theory-on-defeat-of-Spanish-Armada/4979591768000/.

U.S. Department of Energy. n.d. "Fuel Cells." https://www.energy.gov/sites/prod/files/2015/11/f27/fcto_fuel_cells_fact_sheet.pdf.

Venzon, Cliff. 2021. "Philippines and US Boost Defense Ties amid South China Sea Feud." *Nikkei Asia*, April 12, 2021. https://asia.nikkei.com/Politics/International-relations/South-China-Sea/Philippines-and-US-boost-defense-ties-amid-South-China-Sea-feud.

Wayland, Michael. 2021. "GM and Ford Cutting Production at Several North American Plants due to Chip Shortage." *CNBC*. https://www.cnbc.com/2021/04/08/gm-cutting-production-at-several-plants-due-to-chip-shortage.html.

Wen, Yi, and Arias, Maria. 2014. "What Does Money Velocity Tell Us about Low Inflation in the U.S.?" Federal Reserve Bank of St. Louis. https://www.stlouisfed.org/on-the-economy/2014/september/what-does-money-velocity-tell-us-about-low-inflation-in-the-us.

WGAL TV. 2015. "New Equipment Allows Farmers to Pick Produce Laying Down." https://youtu.be/nVYeF_TvtTU.

Wheeler, Donald. 2011. "Problems with Risk Priority Numbers." *Quality Digest*. https://www.qualitydigest.com/inside/quality-insider-article/problems-risk-priority-numbers.html.

Whirlston. n.d. "History of Cotton Harvesting Machine." https://cotton-machine.com/application/History-of-Cotton-Harvesting-Machine.html.

Whittier, George, and Bordelon, Ben. 2022. "The US National Defense Supply Chain Is in Crisis." *Industry Week*, August 3. https://www.industryweek.com/supply-chain/procurement/article/21247914/the-us-national-defense-supply-chain-is-in-crisis.

Wolchover, Natalie. 2012. "How Much Would it Cost to Build the Great Pyramid Today?" *CBS News*, https://www.nbcnews.com/id/wbna46485163.

Zamoyski, Adam. 1987. *The Polish Way*, p. 175. New York: Hippocrene Books.

Index

H

I

J

K

L

M

N

O

P

Q

R

W

Z

Printed in the United States
by Baker & Taylor Publisher Services